HOW TO ESTIMATE CONSTRUCTION COSTS OF ELECTRICAL POWER SUBSTATIONS

HOW TO ESTIMATE CONSTRUCTION COSTS OF ELECTRICAL POWER SUBSTATIONS

JOHN M. BIFULCO

Construction Publishing Company   New York, New York

FIRST EDITION

Library of Congress Catalog Card Number 72-97067

Copyright © Construction Publishing Company, Inc., 1973
All Rights Reserved.

No part of this publication may be reproduced, stored in a retrieval system, or transmitted in any form or by any means without prior written permission of Construction Publishing Company, Inc.

ISBN 0-913634-11-5

Construction Publishing Company, Two Park Avenue, New York, New York 10016.

PREFACE

This book presents a step-by-step method of determining labor costs for erecting electrical power substations. This method will enable estimators, contractors, and engineers who are established in this field to reduce estimating time, and will help professionals and students alike to learn the rudiments of preparing bids for substation construction.

The first step is to determine, through the use of tables provided, the number of man-hours required for the various work categories that make up the job. The itemized man-hour quantities are then multiplied by hourly wage rates to obtain the preliminary labor cost. These hourly rates take into account such burdens as insurance, taxes, welfare, job factor, and inclement weather. To this preliminary labor cost are added overhead, profit, and cost of equipment. The sum is the final labor cost. The factors that affect each component of labor cost are carefully explained so that the reader can make adjustments for local conditions.

The book is divided into five sections and an appendix. Section 1 introduces the reader to a typical power generation and distribution system and describes, with the help of photographs, major substation components. Section 2 presents tabulations of the man-hour values required to install or erect these substation components. These values are averaged from 50 substations in whose construction the author was involved. Section 3 relates the man-hour values to wages and other cost factors and explains how a final cost estimate is derived. Section 4 is similar to Section 2 but contains an expanded list of work elements. Its purpose is to serve as a model record-keeping form so that the user can easily compare cost factors of future jobs with those of completed ones. Section 5 is a compilation of average prices for material and equipment.

In addition to useful reference material, the Appendix contains a sample bid package consisting of specifications, bill of material, and drawings for a 115/69/34.5-kV substation. The reader is expected to use the information in the text to prepare a sample estimate and compare it with one made by the author.

Thanks are due Jules Godel, P.E., for many helpful editorial changes and suggestions. The author also wishes to single out the Pennsylvania Electric Company for their cooperation in providing photographs and other related information. Thanks are due to a number of other utility companies and electrical contractors for their permission to reproduce certain data and figures. Finally, appreciation is expressed to Arthur Tatman of Arthur Tatman and Associates of Cleveland, Ohio, for the photographs of substation structures.

CONTENTS

PREFACE v

INTRODUCTION ix

UNIT TIME FOR SELECTED WORK ELEMENTS 1

MAN-HOUR AND COST ESTIMATES
  FOR SELECTED WORK ELEMENTS 11

COMPLETE LIST OF WORK ELEMENTS
  FOR ESTIMATING COSTS 27

MATERIALS AND EQUIPMENT COSTS 40

APPENDICES
A Abbreviations Used in Text and Tables 69
B Glossary 70
C Sample Bid Package for 115/69/34.5-kV Substations 76
D Graphic Symbols for one-Line Diagrams 95
E Suspension Insular Units 101
F Conductor Spacings 102
G Substation Standards 103

ABOUT THE AUTHOR 105

INTRODUCTION

Power Distribution and Types of Substations

An electrical system, in a broad sense, must include four major categories:

- Generating plants or stations
- Transmission lines
- Substations
- Distribution lines

Electrical power is generated principally from steam turbine units that use fossil or nuclear fuel. Hydroelectric stations play a less significant part in the nation's energy resources. Because large power-generating stations are more efficient, it is more economical to generate energy in large stations and transmit it over long distances.

Ordinarily, electric power is generated at moderate voltage, transmitted at high voltage, and utilized at comparatively low voltage. Voltage may be raised or lowered economically in alternating-current (AC) systems by means of transformers, since these have high efficiencies. Because the weight of conductor necessary to transmit power varies inversely as the square of the transmission voltage, the higher the voltage used in transmission, the less weight required for the cable, i.e., doubling the voltage quarters the cable weight. This saving is offset by the increasing cost of insulators and components with increasing voltage. Future developments in cryogenic superconductor transmission lines will enable long-distance power transmission with minimal losses.

Figure 1 is a block diagram showing the interconnection of the major components of an electrical system. One type of transformer substation steps up the alternating voltage for transmission.

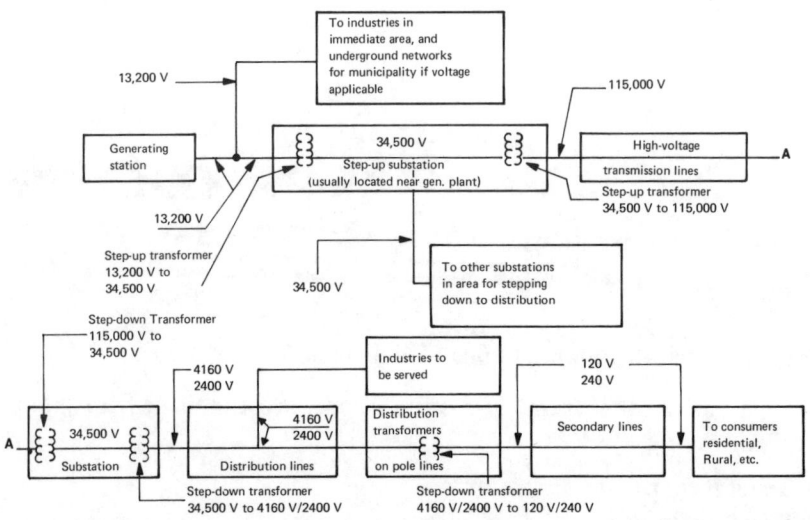

Figure 1. Interconnections of the major components of a typical power system.

Another transformer station steps down voltage for distribution. Switching stations tie two or more supply lines. These are not shown on Figure 1 but are similar to transformer stations. Electrical system designs vary depending upon the size of the area served and the specific use of the power. The transformer stations described in Appendix C (115/69/34.5-kV) and in Section 3 (230/115 kV) are examples of installations in use today.

A perspective view of a switching substation is shown in Figure 2. It is typical of the low-silhouette structures described below, under "Substation Structures." We can see how two feeder lines are tied into a transmission line at the lower right in the picture.

This thumbnail sketch of an electrical system can be augmented by referring to the following texts:

- *Electrical Transmission and Distribution*, Bernhardt Skrotzki. New York, McGraw-Hill Book Company, Inc., 1954.
- *Rural Electrification Engineering*, Unus F. Earp. New York, McGraw-Hill Book Company, Inc., 1950.
- *Electric System Operation*, Bernhardt Skrotzki. New York, McGraw-Hill Book Company, Inc., 1954.

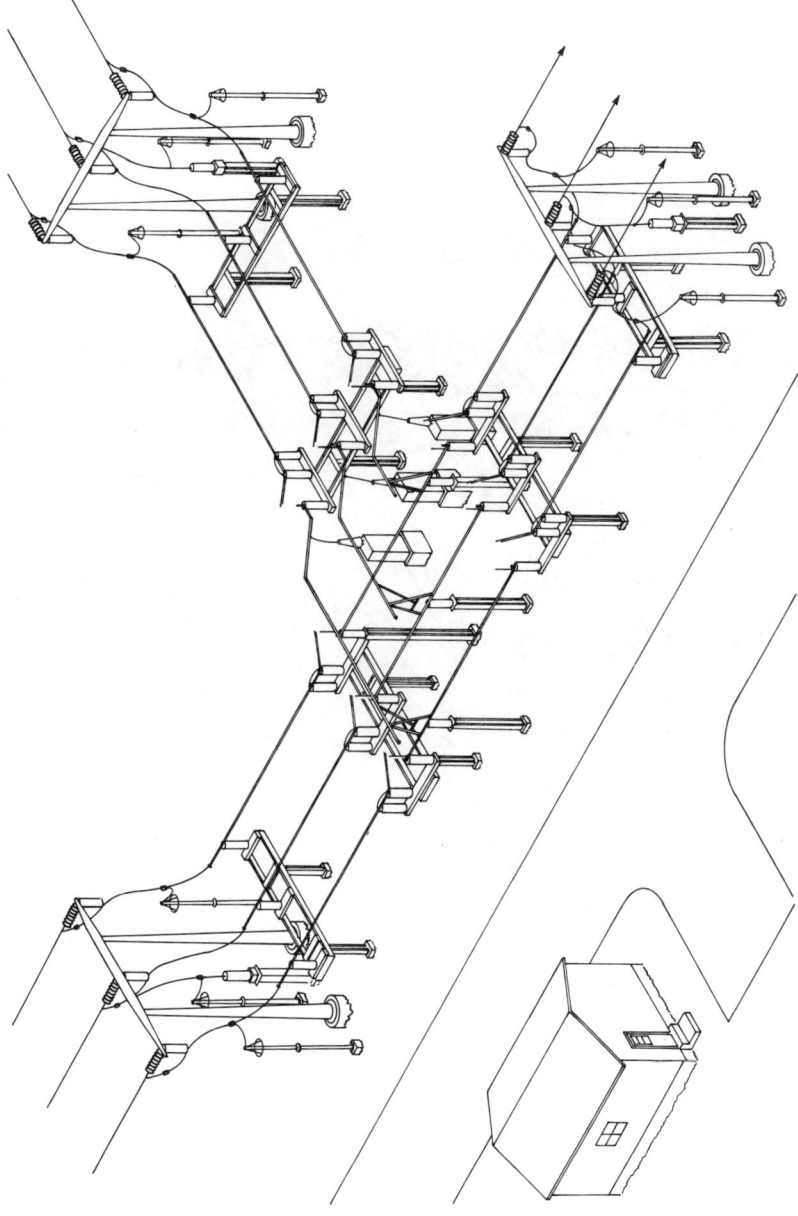

Figure 2. Perspective view of a low-profile switching station, 115 kV.

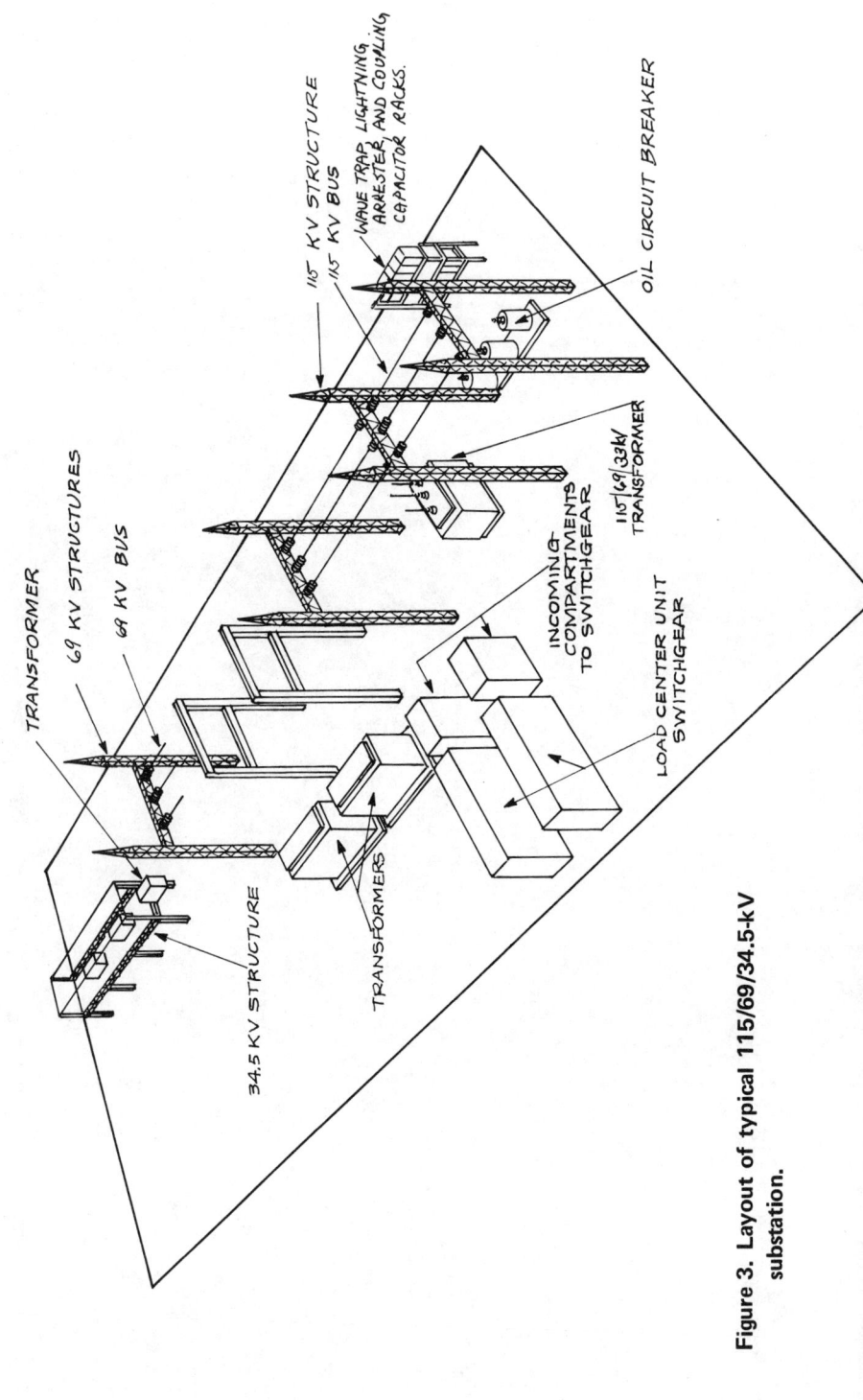

Figure 3. Layout of typical 115/69/34.5-kV substation.

Major Substation Components

The arrangement of equipment varies according to station size, type, and location. It is important that the layout be as uncomplicated as possible; yet consideration must be given to such matters as reliability of operation, safety of personnel, ease of maintenance, and limitation of damage from fire, lightning, or equipment malfunction. The major types of substation equipment can be categorized as follows:

- Transformers
- High-voltage equipment
- Low-voltage equipment
- Relays and instruments

Many components are shown in Figure 3, a layout of a 115/69/34.5-kV transformer substation. Details of this station are provided in Appendix C.

Transformers are usually immersed in oil and cooled by forced oil or water. Various classifications of transformers and other substation components are defined in Appendix B. High-voltage equipment includes lightning arresters, buses, disconnecting and horn gap switches, circuit breakers, and switch gear.

Photographs of some installed equipment are shown in Figures 4 through 7. A 230-kV, oil-cooled transformer with a fan-cooled, oil-to-air heat exchanger is pictured in Figure 4. Note the bus, insulators, and support structure. Circuit breakers with their disconnect switches are shown on a 115-kV substation in Figure 5. A 34.5-kV breaker and disconnects can be seen in detail in Figure 6, and load break switches of the same rating appear in Figure 7.

Low-voltage devices are similar in function to those of higher voltage, but are sized for lighter service requirements. Relays are protective devices that actuate circuit breakers to limit damage to equipment in the event of a short circuit or other abnormal occurrence. Meters and other station instruments are usually mounted on panels in the control house.

With high current and voltage, especially under faulty conditions, the problem in circuit-opening devices is the arcing of contacts. Some circuit breakers are designed with their contacts immersed in oil. The arc heat causes a high-pressure hydrogen bubble

Figure 4. A 230-kV transformer.

Figure 5. Circuit breakers and disconnects in a 115-kV installation.

Figure 6. Circuit breaker and disconnects in a 34.5-kV installation.

Figure 7. Load-break switches in a 34.5-kV installation.

which expands and breaks the arc. In another type, an air blast circuit breaker, high-pressure air is directed at the arc to break it.

Lightning arresters are used in substations to relieve line surges from lightning, switching, or short circuits that would otherwise damage insulators or insulation. Arresters provide a low-resistance path to ground for these surges, and like fuses, interrupt the flow of current as the alternating-current wave passes through zero. In addition to lightning arresters, connections for transformer and other power equipment neutrals must be grounded for the protection of personnel and equipment.

High-voltage buses are made of copper bar or tubing and are supported by insulators. When the potential of a conductor exceeds the dielectric strength of the surrounding air, arcing occurs. Thus, some minimum distance between conductors must be maintained to prevent corona effects and the associated power losses. This distance is dependent upon line voltage. For overhead, flexible conductors, the separation should be great enough to prevent contact due to wind effects. Table 24 in Appendix F is a guide of recommended distance between conductors.

Insulators, usually made of glazed porcelain or Pyrex, isolate buses and overhead conductors from their grounded supports. They must withstand high mechanical stress, resist flashover, and resist weathering. Table 23 in Appendix E gives the average number of suspension insulators used per string for various line voltages.

Most good substation designs allow for future growth. Ample kilovolt-ampere reserve is usually built into the plans and often the placement of future components is shown on the drawings. In preparing a bid, the contractor should be certain he fully understands the scope of the job so he does not inadvertently include the costs for future work in his estimate.

Substation Structures

Although some substation structures are made of aluminum and reinforced concrete, most are constructed of steel. These can be classified by their use, such as supports for line terminals, buses, disconnect switches, and lightning arresters. Circuit breakers and transformers are usually mounted upon concrete foundations.

Often the above-mentioned line terminal, or line dead-end structure as it is sometimes called, is designed as a unit structure or as a single- or double-bay structure. The latter is used in combination with the support of other components such as circuit breakers, feeders, and switches.

Sanderson classifies substation structures as unit type, truss type, and ground or flat type.* The unit type consists of individual towers or A-frame supports that are inexpensive and offer the most flexible arrangement. The truss type is an assembly of beams or lattice trusses interconnected to form a rigid and strong framework for supporting equipment and incoming lines. In the ground or flat type all equipment except the tower to support incoming lines is supported from below and is close to the ground. The absence of overhead steel reduces the likelihood of electrical accidents and generally conforms with the aesthetic advantages of low-profile installations. Another advantage is that components are more accessible for inspection and repair. However, this layout requires more ground space than others.

Figures 8 through 13 depict examples of substation structures. Figure 8 shows a tapered column main structure for a 115-kV station and Figure 9 a straight column main structure for a 69-kV station. Note the air break switches at the top and the three lightning arresters above the lower girder. Figure 10 is a truss-type, single-bay main structure with a separate regulator structure in the foreground. The step voltage regulators are shown in greater detail in Figure 11. Figure 12 shows a 34.5-kV transformer structure. The structures in Figure 13 are more detailed views of those in Figure 5 and are good examples of ground or flat type structures.

The National Electrical Manufacturers Association (NEMA) has established standards (NEMA Standards, Part 36, SG 6-36.01 to 36.09) for outdoor station structures that specify design criteria for loading from apparatus, wind, and ice. The standards also note that since these steel structures are usually galvanized, a high degree of accuracy in fabrication is required since reworking errors in the field removes the protective coating.

*Electric System Handbook, C. H. Sanderson, McGraw-Hill Book Co., New York, 1930.

Figure 8. Tapered column main structure for a 115-kV substation.

Figure 9. Straight column main structure for a 69-kV substation.

Figure 10. Truss-type, single-bay main substation structure with separate regulatory structure.

Figure 11. Detail of voltage regulators in installation shown in Figure 10.

Figure 12. Structure for a 34.5-kV transformer.

Figure 13. Ground-type substation structure.

UNIT TIME FOR SELECTED WORK ELEMENTS

Tables 1 through 6 indicate the average man-hours of labor required for various work categories involved in erecting an electrical power substation. Unit labor time is not given for such items as civil and architectural details, since this book is concerned with the installation of the main structure, supporting stands, equipment racks, and other electrical components. Section 4 contains a complete list of work elements.

Two principal labor classifications are involved in building a substation. Men who erect or work on transmission lines and structures are called *linemen.* In this category are journeymen, foremen, and truck drivers. An *insideman* installs and connects electrical devices such as control switches, relays, meters, cables, panelboards, and lighting. This work, too, is divided between the journeymen who do the work and the foremen who supervise the work. In small crews it is not uncommon for the foreman to work along with his men.

In Section 3 examples are given of the use of these tables. By following through the steps and explanation, the total cost for labor, overhead, and profit can be determined.

The glossary (Appendix B) provides definitions of many components mentioned in the tables so that the reader with limited knowledge can better understand the vocabulary of this highly specialized field.

2 ELECTRICAL POWER SUBSTATIONS

TABLE 1. Unit Time for Erecting Steel Structures

Entries in Table 1 are for the lineman labor classification only.

Voltage (kV)	Item	Average Weight (lb)	Average man-hours per 1,000 lb.	per 100'	Each
	Sort and inventory all steel and switch gear		0.92		
230	Main structure	27,677	7.5		
	3-phase bus support stand (30' to 40' high)	1,722	13.1		
	3-phase bus support stand (20' high)	1,321	8.7		
	1-phase bus support stand	450	4.1		
	Switch rack (25' high)	4,090	8.3		
	Switch rack (13' 6" high)	2,750	9.7		
	Transformer structure	11,500	8.7		
	230/46-kV transformer structure	15,268	8.4		
	Bus potential transformer rack	550	19.1		
	Coupling capacitor potential device rack	370	9.6		
	Lightning arrester rack	773	8.7		
	Tie-down stand	436	10.5		
	Switch base, structural steel	---			1.71
	Wave trap rack	---	8.4		
	Dead-end tower	29,500	7.95		
115	Main structure	23,800	8.6		
	1-phase bus support stand	152	8.2		
	3-phase disconnect switch rack (30' high)	1,638	7.7		
	3-phase disconnect switch rack (15' high)	1,234	5.5		
	Transformer structure	---	8.7		
	115/23-kV transformer structure	5,095	9.0		
	Bus potential transformer rack	590	8.0		
	Coupling capacitor potential device rack	---	9.6		
	Lightning arrester rack	212	6.9		
	Switch base, structural steel	---	---		1.71
	Wave trap rack	---	8.4		
	Dead-end structure	11,050	9.27		

UNIT TIME FOR SELECTED WORK ELEMENTS

TABLE 1. Continued

Voltage (kV)	Item	Average Weight (lb)	Average man-hours per 1,000 lb	per 100'	Each
46	Structure	18,391	8.3		
	Potential transformer rack	---	8.0		
	Lightning arrester rack	---	6.9		
	Aluminum wave trap rack	---			2.0
	Aluminum coupling capacitor rack	---			3.8
34.5	Structure	---	9.0		
	Potential transformer rack	---	8.0		
	Lightning arrester rack	---	6.9		
	Dead-end and switch structure	3,100	10.5		
23	Structure	---	9.0		
	Potential transformer rack	---	18.4		
	Lightning arrester rack	---	6.9		
15	Structure	6,250	8.5		
	3-phase bus support stand (lattice)	956	8.0		
	3-phase bus support stand (I-beam)	383	2.0		
	1-phase potential transformer rack	347	10.0		
	Standard station service transformer rack	253	11.0		
	Standard station emergency service transformer rack	253	11.0		
	Switch operator platform	100	16.5		
	Lightning mast (80' high)	1,643	1.52		
	Dead-end structure (30' high tubular steel)	10,000	5.0		

4 ELECTRICAL POWER SUBSTATIONS

TABLE 2. Unit Time for Installing and Adjusting Disconnect Switches

Entries in Table 2 are for the lineman labor classification only.

Voltage (kV)	Item	Amperes	Average man-hours per 1,000 lb	per 100'	Each
230	Joslyn, type DA	2,000			80.0
	Type RF-2, 3 pole, motor operator	1,200			50.0
	ITE, type TTT, 3 pole, horizontal break, gang operated	1,200			120.0
	SS, type EV 230-1600, 3 pole, gang operated	1,600			70.0
	SS, type EV 230-1600, 3 pole, gang operated with ground switch	1,600			80.0
	Gas blast switch	1,200			85.5
	ITE, type TTR-6, 3 pole with ground switch	1,200			75.0
	ITE, type TTR-6, 3 pole	1,200			56.0
115	ITE, type TTR-6 or TTR-49, 3 pole, gang operated	1,200			43.5
	ITE, type A7, 3 pole	1,200			92.0
	SS, type WAG	1,200			55.0
	ITE, type TTR-6, 3 pole, motor operator	600			48.0
	ITE, type TTR-49	600			40.0
	SS, type WAG, motor operator	600			68.0
	Joslyn, type FB 4 or DN	600			45.0
	SS, type 57-L, 3 pole, gang operated	1,200			60.0
46	ITE, type TTR-49, with motor operator	600			30.5
	SS, type BPO, 1 pole	1,200			1.42
34.5	ITE, type HPL, 1 pole	1,200			1.42
	SS, type EV 34600, 3 pole	600			24.5
12.47	ITE, type HPL, 1 pole	1,200			2.5
	1 pole with 200-A cutout	600			1.6
	1 pole with ground switch	600			1.0
	Motor-operated switch mechanism				10.6
	1-pole fused switch, type SM-5, SS, type BBA				1.94

Manufacturer's designations: Joslyn (Joslyn Mfg. and Supply Co., Chicago, Ill.); ITE (ITE Imperial Corporation, Pittsburgh, Pa.); SS (Southern States, Inc., Hampton, Ga.); S & C (S&C Electric Company, Chicago, Ill.)

TABLE 3. Unit Time for Installing and Terminating Buswork, Insulators, and Connectors

Entries in Table 3 are for the lineman labor classification only.

Voltage (kV)	Item	Average man-hours Per 1,000 lb	Per 100'	Each
230	Insulator stack			5.8
	String, 10" strain insulators			3.1
	Bus supports			4.0
	Conductor tie-down assembly			5.3
	Bushing on potential transformer			4.0
	Wire guide			1.35
	Potential transformer emergency station service			15.0
115	String, 10" insulators			1.66
	Bus supports			2.3
	Wire guide			1.35
34.5	String, 10" insulators			1.14
	Bus supports			1.28
15	Bus supports			2.1
	Bus—all aluminum conductor (AAC)			
	1033.5 MCM AAC		16.3	
	4/0 AAC		16.0	
	Bus—Aluminum conductor, steel reinforced (ACSR)		20.0	
	2493 MCM ACSR		10.0	
	1033.5 MCM ACSR		13.5	
	636 MCM ACSR			
	Bus—aluminum pipe (IPS)			
	5" diam.		20.5	
	4" diam.		12.8	
	2½" diam.		9.9	
	2" diam.		8.0	
	1¼" diam.		6.5	

6 ELECTRICAL POWER SUBSTATIONS

TABLE 3. Continued

Voltage (kV)	Item	Average man-hours Per 1,000 lb	Per 100'	Each
	2" IPS aluminum A–frame assembly			7.0
	Bus—copper pipe (IPS)			
	2½" diam.		16.6	
	2" diam.		9.1	
	1½" diam.		7.8	
	1¼" diam.		10.0	
	1" diam.		7.2	
	¾" diam.		17.0	
	1,000 MCM copper bus		10.2	
	500 MCM copper bus		11.0	
	4/0 copper bus		11.2	
	Bus connections—expansion bus, line and ground *(no welded connections)*			53 per 100
	Bus connections—(welded)—1¼", 2½", 4", 5", etc. IPS Aluminum or copper to Anderson-type connectors			182 per 100
	Static wire			
	#6 Copperweld (CWD)		5.3	
	3/8" extra high strength (EHS) Bethlehem "C"		7.0	
	1/2" EHS Bethlehem "C"		8.0	
	3 #7 Alumoweld		5.6	

TABLE 4. Unit Time for Installing Electrical and Miscellaneous Equipment

Entries in Table 4 are for the lineman labor classification only.

Voltage (kV)	Item	Average man-hours Per 1,000 lb	Per 100'	Each
198	Lightning arrester (including grading rings)			8.7
120	Lightning arrester			3.6
	Lightning arrester (including grading rings)			3.3
99	Lightning arrester			4.7
50	Lightning arrester			1.7
37	Lightning arrester			1.67
12	Lightning arrester			1.8
9	Lightning arrester			1.0
230	Coupling capacitor potential device (CCPD)			8.5
	Wave trap and CCPD			18.5
	Wave trap			8.0
	Line tuner for coupling capacitor			5.5
115	Coupling capacitor potential device			3.7
	Wave trap and CCPD			17.5
46	Coupling capacitor			4.0
	400-A line trap			5.3
230	Potential transformer			12.0
115	Potential transformer			4.85
46	Potential transformer			2.2
34.5	Potential transformer			3.7
12.4	Potential transformer			2.75

8 ELECTRICAL POWER SUBSTATIONS

TABLE 4. Continued

Voltage (kV)	Item	Average man-hours Per 1,000 lb	Per 100'	Each
230	Oil circuit breaker (add for bushings and current transformers but not for installation of breaker)			50.0
	Oil circuit breaker			70.0
115	Oil circuit breaker (includes testing)			67.0
	Oil circuit breaker			50.0
46	Oil circuit breaker, unload and place on foundation			13.0
34.5	Oil circuit breaker			5.3
	Oil circuit breaker (includes testing)			15.0
	Accessory equipment transformer, bushings, radiators, fittings, etc.			187.0
	Ground resistor with support structures on 115/12.47kV transformer			32.0
12.47	Current transformer			2.0
	6,000-kVAR capacitor bank with support structures			38.0
	Joslyn Varmaster, 3 pole, 400 A			9.8
14.4	Vacuum recloser, 140 A			6.8
	Line tuner, Westinghouse type JZ-70			2.3

TABLE 5. Unit Time for Installing Control House and Associated Equipment

Item	Average man-hours Per 1,000 lb	Average man-hours Per 100'	Average man-hours Each	Labor class*
Control house (average 20' x 30'), panels, fixtures, conduits, wiring, etc.			127.0	I
230-V battery charger—rack, set of 6-V batteries (For 125-V dc Operation) and battery rack			26.5	I
230-V battery charger and rack			14.0	I
1 set (20) 6-V batteries and rack			14.0	I
Switchborad (30" x 24" x 7" long panels, 15kV through 230 kV), control boards for breakers)			9.0	I
Supervisory control cabinet			11.3	I
Carrier control cabinet			11.0	I
Cable				
12 conductor, #10		2.4		
7 conductor, #10		2.8		
4 conductor, #10		1.03		
4 conductor, #9		1.8		
2 conductor, #10		0.93		
1 conductor, #4/0		5.1		
1 conductor, #2		2.3		
3 conductor, #10		1.25		
4 conductor, #6		2.0		
#8 carrier current cable		3.33		
RG—34U coaxial cable		1.7		
Terminations of above cables for ac lighting and dc control			17.2 per 100	I
Fibertags, stamp and lacquer (for control cable identification)			5.8 per 100	I
10 circuit, dc distribution panel with conduit risers			9.0	I

*I, insideman.

TABLE 6. Unit Time for Installing Conduit, Grounding, and Yard Lighting

Item	Average man-hours Per 1,000 lb	Per 100'	Each	Labor class*
Conduit (labor includes fittings, connectors, cutting, threading, tamping, trenching, and grouting of pull boxes)				
2" diam. PVC		8.4		L
1½" diam. PVC		9.5		L
3" diam. aluminum		21.5		L
2½" diam. aluminum		20.0		L
2" diam. aluminum		10.0		L
1½" diam. aluminum		11.5		L
1" diam. aluminum		22.8		L
½" diam. aluminum		15.0		L
3" diam. galvanized		35.6		L
2" diam. galvanized		16.9		L
1½" diam. galvanized		17.0		L
1" diam. galvanized		26.6		L
2½" x 1'0" conduit stub			1.34	L
2" x 1'0" conduit stub			0.85	L
2" Sealtite conduit and fittings		31.5		L
Mercury light fixture (includes photocell and Luminaire mounting bracket)			4.8	L
Mercury lamps (includes concrete pier and footing, 3" diam. mast and fittings, mounting bracket, photocell and fittings, and internal connections and wiring)			19.6	L
Mercury light fixture (includes concrete pedestal, photocell, and Luminaire mounting bracket)			13.0	L
Yard convenience outlet			2.6	L
3-phase oil filter outlet			4.5	L
Junction box (average 10" size)			2.5	L
General Electric, Par-38 cluster			3.0	L
Grounding (install ground grid, electrical equipment, structure and manhole grounds, including Cadweld connections, ground connectors, trenching, and backfill)	6.7			L
Megger ground grid (record test results)			14.0	I
Name plates and signs (breaker or circuit names)			1.0	L
Haul material (1 load)			10.0	L
Final clean-up			100.0	L
Move equipment and tools (to and from job site)			47.0	L

*I. insideman; L, lineman.

MAN-HOUR AND COST ESTIMATES FOR SELECTED WORK ELEMENTS

The use of Tables 1 through 6 can best be explained by preparing, step by step, a sample cost estimate. First, we determine the man-hours of labor required for each item of a substation bill of material. Then we examine the type of labor used for the tasks involved, so that the different labor rates can be correctly assigned to various portions of the job. To these labor costs are added overhead as a percentage of total labor cost, the cost of additional equipment, and profit. These calculations yield the quote price used for bidding.

The reader is again reminded that this book's purpose is to specify all the components that determine final costs. The dollar values used in the sample estimate are realistic but not immutable. Local or special conditions, differences in labor rates, and differences in material costs must be taken into account by every estimator.

Total Time Requirements for Work Elements in a 230/115-kV Substation

The information in columns 3 and 4 of Tables 7 through 13 (the description of materials and the quantity in terms of weight, number of items, or linear feet) is usually contained in a bill of material provided by the electrical utility company in its invitation to bid. From this information, the contractor must prepare a competitive bid.

For the purposes of our example, we assume that the utility company provides all materials and components, and that the contractor's only responsibility is to install these substation components. However, if the contractor is required to supply material, its cost *plus* overhead and profit must be added to the labor cost.

12 ELECTRICAL POWER SUBSTATIONS

TABLE 7. Total Time for Erecting Steel Structures

Entries in Table 7 are for the lineman labor classification only.

Task No.	No. of units	Description	Total weight No. of items X Linear feet	Unit time* =	Man hours
1		Sort and inventory all steel and switch gear; total weight of all items in Table 7	142,400 x	0.92 / 1,000	131.0
2	2	230-kV main structure @ 30,000 lb each	60,000 x	7.5 / 1,000 =	450.0
3	2	230-kV switch rack (25' high) @ 5,000 lb each	10,000 x	8.3 / 1,000 =	83.0
4	1	230-kV transformer structure @ 12,000 lb	12,000 x	8.7 / 1,000 =	104.5
5	2	230-kV bus potential transformer rack @ 600 lb each	1,200 x	19.1 / 1,000 =	23.0
6	2	230-kV coupling capacitor potential device rack @ 400 lb each	800 x	9.6 / 1,000 =	7.5
7	2	230-kV lightning arrester rack @ 800 lb each	1,600 x	8.7 / 1,000 =	14.0
8	2	230-kV wave trap rack @ 200 lb each	400 x	8.4 / 1,000 =	3.5
9	2	115-kV main structure @ 25,000 lb each	50,000 x	8.6 / 1,000 =	430.0
10	2	115-kV 3-phase disconnect switch rack (30' high) @ 1,700 lb each	3,400 x	7.7 / 1,000 =	26.0
11	2	115-kV bus potential transformer rack @ 600 lb each	1,200 x	8.0 / 1,000 =	9.5

* From Table 1.

Note that calculations are rounded off to the nearest 0.5 man-hour. If instead, they were rounded off to 1-man-hour, the total would be 1,300 or only 1 man-hour difference. Calculations will continue on the basis of the nearest unit man-hour. The estimator can decide how exact his particular needs are and draw his own conclusions.

MAN-HOUR AND COST ESTIMATES FOR SELECTED WORK ELEMENTS 13

TABLE 7. Continued

Task No.	No. of units	Description	Total weight No. of items × Unit time* Linear feet	=	Man hours
12	2	115-kV coupling potential device rack @ 200 lb each	400 × 9.6 / 1,000	=	4.0
13	2	115-kV lightning arrester rack @ 200 lb each	400 × 6.9 / 1,000	=	3.0
14	2	115-kV wave trap rack @ 100 lb each	200 × 8.4 / 1,000	=	1.5
15	1	Standard station service transformer rack @ 300 lb	300 × 11.0 / 1,000	=	3.5
16	1	Standard station emergency service transformer rack @ 300 lb	300 × 11.0 / 1,000	=	3.5
17	2	Switch operator platform @ 100 lb each	200 × 16.5 / 1,000	=	3.5
			Subtotal man-hours		1,301

TABLE 8. Total Time for Installing and Adjusting Disconnect Switches

Entries in Table 8 are for the lineman labor classification only.

Task No.	No. of units	Description	Total weight No. of items × Unit time* Linear feet	man-hours
18	1	230-kV, 3 pole, gang operated, 1600-A, SS EV 230-1600 with ground switch		80
19	1	115-kV, 3 pole, motor operated, 600-A, SS type WAG		68
			Subtotal (man-hours)	148

* From Table 2.

14 ELECTRICAL POWER SUBSTATIONS

TABLE 9. Total Time for Installing and Terminating Buswork, Insulators, Connectors

Entries in Table 9 are for the lineman labor classification only.

Task No.	No. of units	Description	Total weight No. of items x Linear feet		Unit time*	=	man-hours
20	6	230-kV insulator stack	6	x	5.8	=	35
21	6	String of 230-kV 10" strain insulators	6	x	3.1	=	19
22	6	230-kV bus support	6	x	4.0	=	24
23	2	115-kV 10" insulator stack	2	x	1.66	=	3
24	2	115-kV bus support	2	x	2.23	=	5
25	400'	1033.5 MCM AAC bus	400	x	16.3/100	=	65
26	600'	4" IPS aluminum bus	600	x	12.8/100	=	77
27	700'	1000 MCM copper bus	700	x	10.2/100	=	71
28	100	Bus connections—expansion bus, line and ground (no welded connections)	100	x	53.0/100	=	53
29	50	Bus connections (welded)—1¼", 2½", 4", 5" IPS. Aluminum or copper to Anderson-type connectors	50	x	182.0/100	=	91
30	2,000'	3/8" EHS Bethlehem "C" static wire	2,000	x	7.0/100	=	140
			Subtotal (man-hours)				583

* From Table 3.

TABLE 10. Total Time for Installing Electrical and Miscellaneous Equipment

Entries in Table 10 are for the lineman labor classification only.

Task No.	No. of units	Description	Total weight No. of items x Linear feet		Unit time*	=	man-hours
31	6	198-kV lightning arrester (including grading rings)	6	x	8.7	=	52
32	6	99-kV lightning arrester	6	x	4.7	=	28
33	2	230-kV coupling capacitor potential device	2	x	8.5	=	17

* From Table 4.

MAN-HOUR AND COST ESTIMATES FOR SELECTED WORK ELEMENTS 15

TABLE 10. Continued

34	2	230-kV wave trap	2 × 8.0	=	16
35	2	115-kV coupling capacitor potential device	2 × 3.7	=	7
36	2	115-kV wave trap	2 × 17.5	=	35
37	1	230-kV potential transformer			12
38	1	115-kV potential transformer			5
39	1	230-kV oil circuit breaker			70
40	1	115-kV oil circuit breaker			50
41	1	50-kVA station service transformer			13
42	1	50-kVA emergency station service transformer			13
			Subtotal (man-hours)		318

TABLE 11. Total Time for Installing Control House and Associated Equipment

Entries in Table 11 are for the insideman labor classification only.

Task No.	No. of Units	Description	Total weight No. of items × Linear feet	Unit time*	=	man-hours
43	1	Control house (average size 20' × 20'), install ac and dc panels, fixtures, wiring, etc.				127
44	1	230-kV battery charger-rack, set of 6-V batteries for 125-V dc operation, and battery rack				27
45	4	Switchboards (30" × 24" × 7' panels for breaker control)	4 ×	9.0	=	36
46	1	Supervisory control cabinet				11
47	1	Carrier control cabinet				11
48	2,000'	12 conductor, #10 cable (control)	2,000 ×	2.4/100	=	48
49	500'	2 conductor, #10 cable (control)	500 ×	0.93/100	=	5
50	500'	#8 carrier current cable	500 ×	3.33/100	=	17
51	200'	RG-34U coaxial cable	200 ×	1.7/100	=	3
52	900	Terminations of above cable (for ac lighting and dc control)	900 ×	17.2/100	=	155
53	100	Fibertags, stamp and lacquer (for control cable identification)	100 ×	5.8/100	=	6
			Sub-total (man-hours)			446

* From Table 5.

16 ELECTRICAL POWER SUBSTATIONS

TABLE 12.   Total Time for Installing Conduit, Grounding and Yard Lighting

Task No.	No. of Units	Description	Total weight No. of items Linear feet	x	Unit time*	=	man-hours	Labor class†
54	100'	2" PVC conduit	100	x	8.4/100	=	8	L
55	200'	3" aluminum conduit	200	x	21.5/100	=	43	L
56	400'	3" galvanized conduit	400	x	35.6/100	=	142	L
57	6	Mercury light fixture (includes concrete pedestal, photocell, and Luminaire mounting bracket)	6	x	13.0	=	78	L
58	10	Yard convenience outlets	10	x	2.6	=	26	L
59	4	3-phase outlet for oil filter	4	x	4.5	=	18	L
60	6	Junction box (average 10" size)	6	x	2.5	=	15	L
61	4,000'	Grounding, bare copper wire (includes ground grid, electrical equipment, structure, manhole grounds, Cadweld connections, ground connections, trenching, and backfill)	4,000	x	6.7/100	=	268	L
62		Megger ground grid (record test results)					14	I
63	20	Name plates and signs	20	x	1.0		20	L
64		Haul material (4 loads)	4	x	10.0		40	L
65		Final clean-up					100	L
66		Move equipment and tools (to and from job site)					47	L
					Sub-total (man-hours)		819	

* From Table 6.
† I, insideman; L, lineman.

TABLE 13.   Total Time for All Installations

Installations	Time (man-hours)
Steel structures	1,301
Disconnect switches	148
Buswork, insulators, connectors	583
Electrical and miscellaneous	318
Control house and associated equipment	446
Conduit, grounding, yard lighting	819
Total	3,615

Labor Classification and Rates

The labor classification noted in Tables 7 through 12 shows that linemen account for 3,155 man-hours and insidemen for 460 man-hours. If we assume that the contract requires the job to be completed in 600 hours, it follows from a simple calculation that at least six men are needed to complete the work in the allotted time.

$$\frac{3,615 \text{ man-hours}}{600 \text{ hours}} = 6.03 \text{ men}$$

However, this minimum does not allow for sickness, accidents, or bad weather. It is a matter of individual judgment on the part of the contractor to determine how many more men should augment the minimum crew. For our example, we will use a ten-man crew composed of seven linemen (five journeymen, one foreman, and one truckdriver) and three insidemen (two journeymen and one foreman). This is about a 60% safety factor over the minimum crew and should insure completion of work on schedule without overtime.

Having determined the number of people and their job classifications, we can now begin estimating average labor rates. It is assumed that the reader will familiarize himself with current rates in his own geographical area. The hourly wages shown below were typical in the northeast during 1972-1973 and are meant as illustration only.

Linemen	
Journeymen	$ 5.29
Foreman	5.70
Truckdriver	4.05
Insidemen	
Journeymen	5.55
Foreman	6.11

For our crew, the average wage per hour can be calculated as follows:

Linemen
5 journeymen	@ $ 5.29	= $ 26.45
1 foreman	@ 5.70	= 5.70
1 truckdriver	@ 4.05	= 4.05
		$ 36.20
$ 36.20/7	= $ 5.17 average wage per hour	

Insidemen
2 journeymen	@ $ 5.55	= $ 11.10
1 foreman	@ 6.11	= 6.11
		$17.21
$ 17.21/3	= $ 5.74 average wage per hour	

To this basic hourly wage the contractor, as an employer, must add certain other fixed costs:

Linemen
 $ 5.17 Base average wage per hour
  0.65 Insurance and taxes
  0.10 Welfare
  0.45 Job factor and inclement weather
 $ 6.37

Insidemen
 $ 5.74 Base average wage per hour
  0.23 Mileage allowance
  0.65 Insurance and taxes
  0.27 Job factor
 $ 6.89

For estimating purposes, we round off these figures to $6.40 per hour for outside work and $6.90 per hour for inside work.

Insurance and taxes include local, state, and federal taxes and Social Security (FICA) payments. *Welfare* is the employer's contribution to a union fund. *Job factor* is a multiplier used to adjust labor costs to suit a particular job. For instance, size of project, working conditions, and potential hazard are some allowances used to vary the basic wage rate. The reader should bear in mind that such multipliers can establish only an approximate adjustment. *Inclement weather* affects outside work and must be considered. *Mileage allowance* is arranged with insidemen to cover cost of travel to and from certain job sites.

Although insurance and taxes may change periodically, their rates are well known. Welfare costs are negotiated with a labor union and also are easily available to the estimator. Job factor, inclement weather, and mileage allowance are more variable components that should be carefully evaluated for each job under consideration.

Overhead Costs

We have determined the labor costs to be $6.40 for outside and $6.90 for inside work. The next step is to ascertain the overhead burden and express this as a percentage of labor cost.

Recalling that 600 hours (or 15 weeks) was specified as the time to completion, we list below all the items that normally go into the calculation of overhead. Keep in mind that the actual hourly rates or rental fees are for guidance and should be brought into line with the estimator's local conditions.

Superintendent: 600 hr @ $6.25 per hr	$3,750
Timekeeper: 600 hr @ $3.50 per hr	2,100
Insurance and taxes on above (12% of $5,850)	702
Supervisor	400
Superintendent's expenses	525
Trailer rental (4 months @ $50.00 per month)	200
Pick-up truck	900
Telephone	75
Pre-conference on job, meetings	200
Tools and equipment replacement	350
Tools and equipment to and from job	300
Total overhead	$9,502

The supervisor charge is for part-time supervision which amounts to 3% to 5% of the total overhead. Superintendent's expenses include travel costs to and from job, meals, and hotel costs if the job site is located far from the home office. Other overhead costs are self-explanatory.

It was stated previously that the total man-hours in each work category are:
Outside work = 3,155
Inside work = 460

Therefore, labor costs can be computed:

Outside work 3,155 × $6.40 per hour	=	$20,192.00
Inside work 460 × $6.90	=	3,174.00
Total labor cost		$23,366.00

Overhead, as a percentage of total labor cost, is equal to:

$$\frac{9,502}{23,366} \times 100 = 40.66 \text{ or } 41\%$$

to this add a percentage of
home office overhead   + 6

Total overhead   47%

Overhead percentage varies for each job according to its size, duration, location, and nature of the work to be performed. In our case, the 47% is somewhat high; the average range is between 39% and 43%.

Total Estimated Costs

The total cost is the sum of the unit labor cost plus overhead and profit. To this should be added the cost of heavy construction equipment. If the contractor owns the equipment, its cost may be part of the fixed overhead. Usually, this equipment is expensive and specialized so that many contractors choose to rent it.

Profit is set at 10% of the sum of the labor cost and overhead. This is an average mark-up by the contractor and is strongly dependent upon his competition. It is the only factor over which he has some control, since labor and overhead costs are fixed by outside conditions.

On the basis of the total time determinations shown in Tables 7 through 12, we can calculate the quotation prices of the various components. Table 14 illustrates the calculations for tasks 1 through 4.

TABLE 14. Calculation of Quotation Prices for Tasks 1 Through 4

Task No.	Man-hours	Description and calculation

1	131	Sort and inventory all steel and switch gear
		131 @ $6.40 (rate for outside work)
		= $ 838.40
		    394.00 (47% overhead)
		$1,232.40
		    123.24 (10% profit)
		$1,355.64 (total unit cost)
		Quote $1,360

2	450	230-kV main structure
		450 X $6.40 = $ 2,880
		    1,350 (47% overhead)
		$ 4,230
		    423 (10% profit)
		$ 4,653
		Quote $4,660

3	83	230-kV switch racks
		83 X $6.40 = $ 530
		    250 (47% overhead)
		$ 780
		    78 (10% profit)
		$ 858 (total unit cost)
		Quote $ 860

4	105	230-kV transformer structure
		105 X $6.40 = $ 672
		    316 (47% overhead)
		$ 988
		    99 (10% profit)
		$1,087 (total unit cost)
		Quote $1,090

The 66 tasks that make up the bill of material for the 230/115-kV substation are listed in Table 15, with all the elements that determine their quotation prices. Quotation prices are shown as calculated; in the examples in Table 14 they are rounded off to the nearest $5 increment.

22 ELECTRICAL POWER SUBSTATIONS

TABLE 15. Determination of Quotation Prices for Components of 230/115kV Substation

Task No.	No. of units	Description	Man-hours	Labor cost	Overhead (47%)	Profit (10%)	Quotation price
1		Sort and inventory steel and switch gear	131	$1,838.40	$ 394.05	$123.24	$ 1,355.69
2	2	230-kV main structure	450	2,880.00	1,353.60	423.36	4,656.96
3	2	230-kV switch rack	83	531.20	249.66	78.09	858.95
4	1	230-kV transformer structure	105	672.00	315.84	98.78	1,086.62
5	2	230-kV bus potential transformer rack	23	147.20	69.18	21.64	238.02
6	2	230-kV coupling capacitor potential device rack	8	51.20	24.06	7.53	82.79
7	2	230-kV lightning arrester rack	14	89.60	42.11	13.17	144.88
8	2	230-kV wave trap rack	4	25.60	12.03	3.76	41.39
9	2	115-kV main structure	430	2,752.00	1,293.44	404.54	4,449.98
10	2	115-kV 3-phase disconnect switch rack	26	166.40	78.21	24.46	269.07
11	2	115-kV bus potential transformer rack	10	64.00	30.08	9.41	103.49
12	2	115-kV coupling capacitor potential device rack	4	25.60	12.03	3.76	41.39
13	2	115-kV lightning arrester rack	3	19.20	9.02	2.82	31.04
14	2	115-kV wave trap rack	2	12.80	6.02	1.88	20.70
15	1	Standard station service transformer rack	4	25.60	12.03	3.76	41.39
16	1	Standard station emergency service transformer rack	4	25.60	12.03	3.76	41.39
17	2	Switch operator platforms	4	25.60	12.03	3.76	41.39
18	1	230-kV, 3 pole, gang operated 1600-A SSEV 230-1600, with ground switch	80	512.02	240.64	75.26	827.90
19	1	115-kV, 3-pole, motor operated 600-A SS-type WAG	68	435.20	204.54	63.97	703.71

MAN-HOUR AND COST ESTIMATES FOR SELECTED WORK ELEMENTS

#	Qty	Description				
20	6	230-kV insulator stack	35	224.00	105.28	32.93	362.21
21	6	230-kV strain 10" strain insulator	19	121.60	57.15	17.88	196.63
22	6	230-kV bus support	24	153.60	72.19	22.58	248.37
23	2	115-kV 10" insulator stack	3	19.02	9.02	2.82	31.04
24	2	115-kV bus support	5	32.00	15.04	4.70	51.74
25	400'	1,033.5 MCM AAC bus	65	416.00	195.51	61.15	672.67
26	600'	4" IPS aluminum bus	77	492.80	231.61	72.44	796.85
27	700'	1,000 MCM copper bus	71	454.40	213.57	66.80	734.77
28	100'	Bus connections (not welded)	53	339.20	159.42	49.86	548.48
29	50	Bus connections (welded)	91	582.40	273.73	85.61	941.74
30	2,000'	3/8" EHS Bethlehem "C" static wire	140	896.00	421.12	131.71	1,448.83
31	6	198-kV lighting arrester	52	332.80	156.42	48.92	538.14
32	6	99-kV lightning arrester	28	179.20	84.22	26.34	289.76
33	2	230-kV coupling capacitor potential device	17	108.80	51.14	15.99	175.93
34	2	230-kV wave trap	16	102.40	48.13	15.05	165.58
35	2	115-kV coupling capacitor potential device	7	44.80	21.06	6.59	72.45
36	2	115-kV wave trap	35	224.00	105.28	32.93	362.21
37	1	230-kV potential transformer	12	76.80	36.10	11.29	124.19
38	1	115-kV potential transformer	5	32.00	15.04	4.70	51.74
39	1	230-kV oil circuit breaker	70	448.00	210.56	65.86	724.42
40	1	115-kV oil circuit breaker	50	320.00	150.40	47.04	517.44
41	1	50-kVA station service transformer	13	83.20	39.10	12.23	134.53
42	1	50-kVA emergency station service transformer	13	83.20	39.10	12.23	134.53
43	1	Control house (install ac and dc panels, fixtures, wiring, etc.)	127	876.30	411.86	128.82	1,416.98
44	1	230-V battery charger	27	186.30	87.56	27.39	301.25
45	1	Switchboards	36	248.40	116.75	36.51	401.66

24 ELECTRICAL POWER SUBSTATIONS

TABLE 15. Continued

Task No.	No. of units	Description	Man-hours	Labor cost	Overhead 47%	Profit (10%)	Quotation price
46	1	Supervisory control cabinet	11	75.90	35.67	11.16	122.73
47	1	Carrier control cabinet	11	75.90	35.67	11.16	122.73
48	2,000'	12 conductor, #10 cable	48	331.20	155.66	48.69	535.55
49	500'	2 conductor, #10 cable	5	34.50	16.22	5.07	55.79
50	500'	#8 carrier current cable	17	100.30	47.14	14.74	162.18
51	200'	RG-34U coaxial cable	3	20.70	9.73	3.04	33.47
52	900	Terminations, cable	155	1,069.50	502.67	157.22	1,729.39
53	100	Fibertags	6	41.40	19.46	7.07	67.93
54	100'	2" PVC conduit	8	51.20	24.06	7.53	82.79
55	200'	3" aluminum conduit	43	275.20	129.34	40.45	444.99
56	400'	3" galvanized conduit	142	908.80	427.14	133.59	1,469.53
57	6	Mercury light fixture	78	499.20	234.62	73.38	807.20
58	10	Yard convenience outlets	26	166.40	78.21	24.46	269.07
59	4	3-phase outlet for oil filter	18	115.20	54.14	16.93	186.27
60	6	Junction box	15	96.00	45.12	14.11	155.23
61	4,000'	Grounding, bare copper wire	268	1,715.20	806.14	252.13	2,773.47
62		Megger ground grid	14	96.60	45.40	14.20	156.20
63	20	Name plates and signs	20	128.00	60.16	18.82	206.98
64		Haul material (4 loads)	40	256.00	120.32	36.63	413.95
65		Final clean-up	100	640.00	300.80	94.08	1,034.88
66		Move equipment and tools (to and from job site)	47	300.80	141.38	44.21	486.39
		Total quotation price					$37,797.61

MAN-HOUR AND COST ESTIMATES FOR SELECTED WORK ELEMENTS

The only additional cost that must be considered to obtain bid price is that of special vehicles and equipment. We do not attempt to estimate this cost, since contractors may use different methods of doing similar work. Typical costs for renting equipment are shown in Table 16. It is left for the reader to determine the length of time particular vehicles will be in use.

TABLE 16. Typical Costs for Renting Equipment

Vehicles and equipment	Rental costs		
	Month	Week	Hour
Hydraulic truck	$900	$300	$12
Skyhook truck	750	250	10
Pick-up truck	200	50	2
Line truck	250	80	4
Trencher			8
Oil filter			7

There is a useful economic benchmark in substation estimating. Just as some construction is evaluated on the basis of cost per square foot of occupied space, in this field the cost of the substation divided by the number of man-hours on the job gives a similar value. The approximate factor is $12, but is expected to rise to $13 owing to increasing construction costs. Thus, if the total man-hours required to do the job are known, a rapid preliminary estimate of costs can be made. In our example we found that 3,615 man-hours were needed. Thus 3,615 man-hours × $12 = $43,380. Although this is about 14% more than the actual estimate ($37,800), it is close enough for a "ballpark" estimate and is a useful check for gross errors in the final estimate. The error would be reduced if the cost of vehicles and equipment were added to the $37,800.

Extra Work

Extra work is work that is not mentioned in the contract but which the customer wants completed while the contractor is still on the job. Examples are removing old buswork, pulling additional cables, and moving (or removing) existing circuit breakers and insulators.

A conventional arrangement for payment for extra work is based upon cost of labor and materials supplied by the contractor plus 10% profit. The cost of labor is usually the base (or bare) labor rate ($5.29 per hour was the base rate for a journeyman-lineman in the example on page 17), plus some aggregate percentage of the base rate taken from Table 17.

TABLE 17. Computation of Percentage of Base Labor Rate

Item	Percent of base labor rate
Insurance	0.4
Social Security	4.4
Administrative overhead	5.0
Workmen's Compensation	0.9
Other (specify)	
Taxes	3.7
NECA	1.0
Apprentice training	1.0
Total	16.4

Adding 16.4% to the base rate gives $6.16 per hour. Then $0.10 per hour for union health and welfare benefits can be included if applicable. For our case, the contractor may charge $6.26 per hour plus materials, plus a 10% profit on labor and materials.

COMPLETE LIST OF WORK ELEMENTS FOR ESTIMATING COSTS

This section contains a master list of work elements and a suggested form for computing the estimate. The purpose of the master list is to alert the contractor to the various tasks involved.

Since substations are of varied capacities, purposes, and locations, components differ from station to station. Therefore, no attempt is made to present a standard list of work elements to suit a particular substation. Instead, the author has reviewed the estimates of more than 50 installations and has compiled a step-by-step breakdown of all work that might be encountered. It is left to the contractor to pick out those elements that apply to his specific job and transfer them to the computation form.

Section 2 was concerned with selected work elements and the man-hours required for their completion. Civil, architectural, and other nonelectrical details were not covered because it is not the primary purpose of this book to supply this information. However, the contractor may be required to provide these services and should be prepared to include them in his bid. A recommended source of cost information for civil and mechanical construction operations is the *Building Cost File* (New York, Construction Publishing Company, Inc.) The *File* is available in four regional issues and is updated each year.

The master list and computation form do *not* take into account the cost of material. If materials are supplied by the contractor, their costs, plus overhead and profit, must be added to the labor costs.

28 ELECTRICAL POWER SUBSTATIONS

Complete List of Substation Work Elements

YARD
- Yard layout engineering
- Grading (cut and fill)
- Seed slopes
- Chemical treatment
- Install fence (including piers)
- Concrete trench
    - Excavate
    - Form and reinforce
    - Pour concrete
    - Backfill
- Duct runs
- Conduit
- Junction/pull boxes
- Manhole
    - Excavate
    - Form and reinforce
    - Pour concrete
    - Backfill
- Install ground grid (include terminations)
- Install ground platforms
- Yard surfacing
- Drill well
- Septic tank and drain field
- Roadway
- Install corner posts
- Install yard lighting
- Landscape lot

CONTROL HOUSE
- Control house foundation
- Foundation (includes cable trench)
- Erect complete building
- Wire control house (including ac and dc panels, lights, meter cabinet, receptacles)
- Install plumbing
- Install toilet enclosures

COMPLETE LIST OF WORK ELEMENTS FOR ESTIMATING COSTS 29

Install heaters
Install battery charger
Install station battery (include rack)
Install switchboards (include cable terminations)
Install miscellaneous furniture
Install supervisory control
Install map board
Install control house insulation and trim

FOUNDATIONS
    230-kV main structure
        Excavate
        Form and reinforce
        Pour concrete
        Backfill
    230-kV bus structures
        Excavate
        Form and reinforce
        Pour concrete
        Backfill
    230-kV switch racks
        Excavate
        Form and reinforce
        Pour concrete
        Backfill
    230-kV potential and lightning arrester racks
        Excavate
        Form and reinforce
        Pour concrete
        Backfill
    230-kV wave trap and coupling capacitor rack
        Excavate
        Form and reinforce
        Pour concrete

    Backfill
230-kV tie-down tower
    Excavate
    Form and reinforce
    Pour concrete
    Backfill
115-kV main structure
    Excavate
    Form and reinforce
    Pour concrete
    Backfill
115-kV transformer structure
    Excavate
    Form and reinforce
    Pour concrete
    Backfill
115-kV lightning arrester and potential transformer racks
    Excavate
    Form and reinforce
    Pour concrete
    Backfill
115-kV wave trap and coupling capacitor
    Excavate
    Form and reinforce
    Pour concrete
    Backfill
115-kV dead-end structure
    Excavate
    Form and reinforce
    Pour concrete
    Backfill
115-kV switch rack
    Excavate
    Form and reinforce
    Pour concrete
    Backfill
46-kV main structure
    Excavate
    Form and reinforce

    Pour concrete
    Backfill
46-kV lightning arrester and potential transformer racks
    Excavate
    Form and reinforce
    Pour concrete
    Backfill
46-kV turning structure
    Excavate
    Form and reinforce
    Pour concrete
    Backfill
34.5 kV main structure
    Excavate
    Form and reinforce
    Pour concrete
    Backfill
34.5-kV lightning arrester and potential transformer racks
    Excavate
    Form and reinforce
    Pour concrete
    Backfill
34.5-kV turning structure
    Excavate
    Form and reinforce
    Pour concrete
    Backfill
23-kV main structure
    Excavate
    Form and reinforce
    Pour concrete
    Backfill
23-kV lightning arrester and potential transformer racks
    Excavate
    Form and reinforce
    Pour concrete
    Backfill
23-kV turning structure
    Excavate

32 ELECTRICAL POWER SUBSTATIONS

        Form and reinforce
        Pour concrete
        Backfill
23-kV power transformer
        Excavate
        Form and reinforce
        Pour concrete
        Backfill
23-kV potential transformer
        Excavate
        Form and reinforce
        Pour concrete
        Backfill
23-kV station service transformer
        Excavate
        Form and reinforce
        Pour concrete
        Backfill
230-kV oil circuit breaker
        Excavate
        Form and reinforce
        Pour concrete
        Backfill
115-kV oil circuit breaker
        Excavate
        Form and reinforce
        Pour concrete
        Backfill
46-kV oil circuit breaker
        Excavate
        Form and reinforce
        Pour concrete
        Backfill
34.5-kV oil circuit breaker
        Excavate
        Form and reinforce
        Pour concrete
        Backfill
23-kV oil circuit breaker

COMPLETE LIST OF WORK ELEMENTS FOR ESTIMATING COSTS

Excavate
Form and reinforce
Pour concrete
Backfill

STRUCTURES
230-kV
Main structure
Bus structure
Switch racks
Transformer structure
Bus potential and lightning arrester racks
Wave trap and coupling capacitor rack
Hold-down tower

115-kV
Main structure
Transformer structure
Lightning arrester and potential racks
Wave trap and coupling capacitor racks
Dead-end structure
Switch rack

46-kV
Main structure
Lightning arrester and potential racks
Turning structure

34.5-kV
Main structure
Lightning arrester and potential racks
Turning structure

23-kV
Main structure
Lightning arrester and potential racks
Turning structure

25-kV main structure
23 and 34.5-kV main structure

115-kV
4-pole switch structure
Unload steel
Sort steel

SWITCHES
Horn gap switch, type TTR
- 115-kV
- 46-kV
- 34.5-kV

Oil circuit breaker disconnect switch
- 230-kV   type TTT
- 115-kV   type TTS
  - type A
- 46-kV   type TTS
  - type A
  - type HPS
- 34.5-kV   type A
  - type HPL

Ground Switch
- 115-kV
  - 3-pole ground switch
  - 1-pole, 12-cycle ground switch
  - 1-pole, 5-cycle ground switch
  - 1-pole, 7-cycle ground switch
  - MO-9 switch operating mechanisms
  - Fuse mounting

BUSWORK
230-kV
- Strain bus conductor and taps
- Tubing bus and taps
- Bus supports and fittings
- Strain insulators

115-kV
- Strain bus conductor and taps
- Tubing bus and taps
- Bus supports and fittings
- Strain insulators

46-kV
- Strain bus conductor and taps
- Tubing bus and taps
- Bus supports and fittings
- Strain insulators

COMPLETE LIST OF WORK ELEMENTS FOR ESTIMATING COSTS 35

34.5-kV
    Strain bus conductor and taps
    Tubing bus and taps
    Bus supports and fittings
    Strain insulators
    Static wire
    Guy wire
COMMUNICATIONS EQUIPMENT
  Install
    400-A wave trap
    800-A wave trap
    1,200-A wave trap
    1,600-A wave trap
    Coupling capacitor
    Turning unit
    Control cable

ELECTRICAL EQUIPMENT
  1-phase and 3-phase power transformers
    Transport transformers and place on foundation
    Haul and filter oil
    Install auxiliary transformer equipment
    Terminate bus and control cable
    Checks and tests
  Potential transformers
    Transport transformers and place on foundation
    Terminate bus and control cable
  Station service transformers
    Transport and place transformers
    Terminate bus and control cable
    Install service control center
    Install outdoor meters
  230-kV oil circuit breakers
    Transport and place on foundation
    Install auxiliary equipment
    Haul and filter oil
    Install bushing potential device
    Install linear couplers
    Install bushing current transformer

Terminate bus and control cable
Checks and tests
Install air compressor
115-kV oil circuit breakers
Transport and place on foundation
Install auxiliary equipment
Haul and filter oil
Install bushing potential device
Install linear couplers
Install bushing current transformer
Terminate bus and control cable
Checks and tests
Install air compressor
46-kV oil circuit breakers
Transport and place on foundation
Install auxiliary equipment
Haul and filter oil
Terminate bus and control cable
Checks and tests
Install air compressor
34.5-kV oil circuit breakers
Same steps as 46-kV oil circuit breaker
23-kV oil circuit breakers
Same steps as 46-kV oil circuit breaker

PROTECTIVE EQUIPMENT
Install   230-kV lightning arresters
          115-kV lightning arresters
          46-kV lightning arresters
          34.5-kV lightning arresters
          23-kV lightning arresters

RELAY AND AUXILIARY EQUIPMENT
Provide checks and tests
Note: The utility company usually will make relay tests and checks, and sometimes test auxiliary equipment. Specifications should clearly state whether contractor has any responsibility for testing.

COMPLETE LIST OF WORK ELEMENTS FOR ESTIMATING COSTS 37

MISCELLANEOUS
  Install temporary grounds
  Repair existing road and bridges
  Miscellaneous supplies
  Final clean-up
  Unload miscellaneous material
  Install name plates

FACTORS AFFECTING HOURLY LABOR RATE
  Payroll insurance
    Workmen's Compensation
    Public liability
    Property damage
    Other (specify)
  Payroll taxes
    FOAB
    Unemployment
    Other (specify)
  IBEW
    Local employees benefit board
    Apprentice training committee
    Health and welfare fund
    Other (specify)
  Other
    Job factor
    Inclement weather

FACTORS AFFECTING OVERHEAD
  Superintendent
  Clerk
  Timekeeper
  Insurance and taxes (for the above personnel)
  Supervisor
  Travel expenses of superintendent
  Moving equipment to job
  Returning equipment from job
  Warehouse or office rental
  Communications equipment rental (radios)
  Utility rental (telephone, electric power, etc.)

Employee's travel allowance
Miscellaneous

OTHER COMPONENTS
Cost of renting heavy construction equipment
Contractor's profit

Suggested Form for Computation

The form shown as Table 18 is a combination of the forms used for Tables 7 through 16. This combination brings together in one place all the calculations required for a substation cost estimate. By classifying equipment into the categories used in Tables 1 through 6 (steel structures; disconnect switches; buswork, insulators, connectors; electrical and miscellaneous equipment; control house and associated equipment; and conduit, grounding, and yard lighting) and compiling subtotals of costs for each category, the contractor will have a better understanding of the factors that affect his estimate.

The practicing estimator might wish to reproduce the following form on a conveniently sized sheet of paper that could eventually become part of bound records. The accumulation of these records, plus the knowledge of where each estimate was in error, should be invaluable to the professional. The use of these records will enable him to become more competitive in bidding for new work and hopefully will help him avoid making a serious mistake while doing so.

TABLE 18. Summary Form for Computing Substation Costs for Each Work Category

Task No.	No. of units	Description	Total weight No. of items Linear feet	×	Unit time	=	Man-hours	Labor class	Labor cost	Over-head	Profit	Quote price

Subtotal

Method of determining labor costs, overhead charges, and profits is described in Section 3.

MATERIAL AND EQUIPMENT COSTS

Equipment and material for the construction of an electrical substation are usually provided by the utility company. However, there are times when the contractor must supply certain components. For these occasions, especially when the bid is competitive, the contractor must obtain an accurate and firm price from the manufacturer or supplier.

Sometimes a contractor is asked to provide approximate, even "ballpark," estimates for equipment that are subject to verification and negotiation. This can occur on an informal basis at a pre-bid meeting or in the field while work is progressing. Since price information must come from many companies, a virtual library of catalogs would be needed to have this information at hand.

Because this volume is intended as a workbook, approximate cost data for equipment and materials are included as Table 19. The list consists of 24 major subject classifications arranged alphabetically. About 600 components within these classifications are identified and priced. The prices shown are for 1973 and are FOB job site. The blank columns to the right of the 1973 cost column allow price changes to be noted as required. Items are often specified by abbreviations and code nomenclature. While this system is easily intelligible to the experienced substation estimator, the beginner should refer to both the glossary (Appendix B) and list of abbreviations (Appendix A).

The reader is again reminded that the prices below are only approximate.

TABLE 19. Approximate Cost of Materials and Equipment for Electrical Substation

Material and Equipment	Cost	
	1973	Current Price
1 ARMOR RODS (cost per set)		
#4/0 ACSR 6/1	$ 3.15	
266.8 MCM ACSR 6/7	3.55	
336.4 MCM ACSR 26/7	5.65	
477 MCM ACSR 24/7	7.80	
636 MCM ACSR 24/7	9.30	
1033.5 MCM ACSR 54/7	16.70	
2 ARRESTERS, LIGHTNING (cost each)		
Distribution type (valve top gap)		
15 kV	39.10	
12 kV	31.05	
9 kV	17.50	
6 kV	15.05	
3 kV	12.10	
Station type		
138 kV	20.70	
99 kV	14.23	
50 kV	8.05	
37 kV	4.95	
12 kv	2.42	
9 kV	2.23	

TABLE 19. Continued

Material and Equipment		Cost	
		1973	Current Price
3	BUS (cost per 10')		
	Aluminum, IPS size		
	5"	113.85	
	4"	72.45	
	2½"	26.45	
	2"	17.25	
	1¼"	10.35	
	Copper, IPS size (threadless)		
	2½"	32.20	
	2"	24.15	
	1½"	18.40	
	1¼"	15.55	
	1"	12.65	
	¾"	10.35	
4	CABLE, CONDUCTOR, AND SPLICES		
4.1	Bare (cost per 100' feet)		
	#4 solid copper	8.34	
	#1 stranded copper	17.83	
	#2/0 stranded copper	28.75	
	#4/0 stranded copper	49.45	
	1,000 MCM stranded copper	176.00	

MATERIAL AND EQUIPMENT COSTS 43

	#4A Copperweld	9.60
	#6A Copperweld	6.35
	#2, 7/1 ACSR	3.85
	#2/0, 6/1 ACSR	6.90
	#4/0, 6/1 ACSR	11.50
	336.4 MCM, 26/7 ACSR	19.55
	477 MCM, 24/7 ACSR	28.75
	636 MCM, 24/7 ACSR	36.25
	1033.5 MCM, 54/7 ACSR	55.30
4.2	Control cable (600 V) (cost per 100')	
	12 conductor, #10	57.50
	7 conductor, #10 polyethylene	37.40
	7 conductor, #9 polyethylene	46.00
	4 conductor, #10	24.15
	4 conductor, #9	34.50
	3 conductor, #10	18.40
	5 conductor, #10 polyethylene	33.35
4.3	Covered conductor (cost per 100')	
	#4 stranded copper	12.20
	#1 stranded copper	26.70
	#4/0 stranded copper	63.95
	500 MCM stranded copper	155.25
	#4, 7/1 ACSR	4.25
	366.4 MCM stranded aluminum	23.30
4.4	Open cable (cost per 100')	
	3/8" Copperweld, 7 conductor, #8 HS	25.45

TABLE 19. Continued

		Cost	
	Material and Equipment	1973	Current Price
	1/2" Copperweld, 7 conductor, #6 EHS	28.75	
	7 conductor, #9 AWA messenger	9.90	
	7 conductor, #7 AWA messenger	15.55	
	Open cable (12.5 kV)		
	#4/0, 6/1 ACSR polyethylene	16.55	
	336.4 MCM 19 stranded aluminum	31.05	
	Open cable (23 and 34.5 kV)		
	#4/0, 6/1 ACSR polyethylene	37.60	
	336.4 MCM, 26/7 ACSR polyethylene	41.65	
	#4/0 stranded aluminum polyethylene	48.30	
	336.4 MCM 19 stranded aluminum polyethylene	58.65	
4.5	Underground cable (600 V) (cost per 100')		
	#2 stranded copper	29.35	
	#2/0 stranded copper	51.75	
	#4/0 stranded copper	78.20	
	250 MCM stranded copper	106.95	
	500 MCM stranded copper	204.70	
	#4 - 7 stranded aluminum	23.00	
	#2 - 7 stranded aluminum	28.75	
	#2/0 - 19 stranded aluminum	46.25	

MATERIAL AND EQUIPMENT COSTS 45

250 MCM - 19 stranded aluminum	95.45
500 MCM - 37 stranded aluminum	161.00
Underground cable (15 kV) (cost per 100')	
#2 stranded copper, shielded butyl	102.35
#2/0 stranded copper, shielded butyl	119.60
#4/0 stranded copper, shielded butyl	193.20
300 MCM stranded copper, shielded butyl	207.00
300 MCM, 3 conductor paper and lead	488.75
500 MCM stranded copper, shielded butyl	289.80
#2 - 7 stranded aluminum	42.55
#4/0 - 7 stranded aluminum	54.65
500 MCM - 37 stranded aluminum	122.50
Underground cable (23 kV) (cost per 100')	
#1 copper, shielded neoprene, paper and lead	162.15
#1/0 copper, shielded neoprene, paper and lead	135.70
#1 - 19 stranded aluminum	56.35
#4/0 - 19 stranded aluminum	68.10
500 MCM - 37 stranded aluminum	151.80
Underground cable (35 kV) (cost per 100')	
#1 - 19 stranded aluminum	83.95
#4/0 - 19 stranded aluminum	105.80
500 MCM - 37 stranded aluminum	188.60
4.6 Underground splices (cost each)	
#2 - 15 kV	14.95
#4/0 - 15 kV	20.70

TABLE 19. Continued

Material and Equipment	Cost	
	1973	Current Price
500 MCM - 15 kV	33.35	
#1 - 25 kV	21.30	
#4/0 - 25 kV	29.35	
500 MCM - 25 kV	46.00	
#1 - 35 kV	27.60	
#4/0 - 35 kV	37.95	
500 MCM - 35 kV	56.95	
5 CIRCUIT BREAKERS (cost each without discount)		
230-kV OCB, 2,000 A	85,100.	
169-kV OCB, 1,600 A	67,160.	
115-kV OCB, 1,600 A	44,620	
69-kV OCB, 2,000 A	35,190.	
46-kV OCB, 1,200 A	22,425.	
34.5-kV OCB, 1,200 A	12,880.	
14.4-kV OCB, 1,200 A	12,765.	
6 CLAMPS		
6.1 Post insulator clamps (cost each)		
#2 to 4/0 copper	1.75	
#2 to 4/0 ACSR	2.30	
336.4 to 466 MCM ACSR	2.75	
636 MCM ACSR	5.20	

MATERIAL AND EQUIPMENT COSTS 47

6.2	Strain clamps (cost each)	
	#4 to 2/0 stranded copper	2.90
	#6A and #4A Copperweld, #4 to #1/0 ACSR	3.00
	#2 to #4/0 all aluminum, #4 to #4/0 ACSR	3.10
	#1 to #4/0 stranded copper	6.00
	#4/0 to 300 MCM stranded copper, #2 to #4/0 ACSR	10.30
	300 MCM to 500 MCM stranded copper	12.90
	266.8 MCM ACSR	12.20
	336.4 MCM ACSR	3.10
	Strateline — #4/0 copper	5.05
	Strateline — 500 to 750 MCM copper	16.80
	477 MCM ACSR	23.00
	636 MCM ACSR	2.75
	Strateline #2/0 and #4/0 ACSR, #4/0 aluminum	
	Strateline - 336.4 MCM ACSR, 336.4 MCM Aluminum	4.75
	1033.5 MCM ACSR	27.95
6.3	Suspension clamps (cost each)	
	#4 solid to #2/0 stranded copper	2.40
	#3/0 to #4/0 stranded copper, #4 to #2 ACSR	2.45
	#1/0 to #4/0 ACSR	3.70
	250 MCM to 500 MCM copper, 266.8 MCM ACSR	4.45
	336.4 MCM ACSR	5.65
	477 MCM ACSR	6.35
	636 MCM ACSR	7.50
	1033.5 MCM ACSR	10.10

TABLE 19. Continued

Material and Equipment	Cost	
	1973	Current Price
7 CONDUIT AND ACCESSORIES		
Conduit (cost per 10')		
Galvanized, 1"	3.10	
1½"	4.70	
2"	6.35	
3"	12.70	
4"	18.70	
5"	40.10	
Aluminum, ½"	1.60	
1"	2.75	
1½"	4.60	
2"	5.85	
2½"	9.90	
3"	13.25	
4"	20.70	
5"	31.05	
Lightwall plastic, 2"	2.05	
4"	10.35	
5"	12.40	
Heavywall plastic, 6"	39.10	

MATERIAL AND EQUIPMENT COSTS 49

Fiber,	2"	3.10
	4"	7.60
	6"	10.65
Transite,	4"	4.90
	5"	7.60
Polyvinylchloride (PVC),	2"	4.90
	3"	9.45
	4"	15.40
	5"	31.10
Conduit accessories (cost each)		
Adapter		
Galvanized to PVC,	2"	0.80
	3"	2.30
	4"	3.60
Transite to aluminum,	4"	0.85
	5"	1.15
Bushing, plastic,	2"	0.30
	3"	0.60
	4"	0.80
Conduit strap, 2 hole,	2"	0.35
	3"	0.35
	4"	1.10
Conduit strap, galvanized	4"	3.60
	5"	7.55
Ground clamp,	2" conduit	0.86
	3" conduit	3.20

TABLE 19. Continued

Material and Equipment	Cost 1973	Current Price
Galvanized weatherhead, 1½"	2.40	
2"	4.90	
3"	25.20	
4"	41.70	
Plastic bend, 90° long sweep, 2"	2.90	
4"	14.80	
45° long sweep, 4"	15.05	
Fiber bend, 45° long sweep, 4"	7.15	
90° long sweep, 4"	8.85	
Galvanized bend, 90° long sweep, 4"	42.40	
Spacers, 1" separation (cost each)		
4" intermediate	0.35	
4" base	0.60	
5" base	0.35	
8 CONNECTORS		
8.1 Compression, (insulated) (cost each)		
#4 aluminum to #6 copper	0.25	
#4 aluminum to #4 aluminum or #4 copper	0.30	
#4 ACSR to #6 solid copper	0.30	

MATERIAL AND EQUIPMENT COSTS 51

	#2 aluminum to #6 copper	0.25
	#2 aluminum to #4 copper	0.25
	#2 aluminum to #2 copper	0.25
	#2 aluminum to #8 aluminum or #8 copper	0.25
8.2	Compression, (uninsulated cost each)	
	#4 aluminum to #6 copper	0.25
	#4 aluminum to #4 aluminum or #4 copper	0.15
	#2 aluminum to #6 copper	0.15
	#2 aluminum to #4 copper	0.15
	#2 aluminum to #2 copper	0.15
	#2/0 aluminum to #2/0 aluminum	1.30
	#2/0 aluminum to #1/0 copper	1.30
8.3	Splicing sleeve (cost each)	
	7 – #7 Alumoweld	4.60
	#4/0 – 6/1 ACSR	5.20
	336.4 MCM, 26/7 ACSR	6.85
	#4 ACSR – 6/1	1.05
	#2 ACSR – 7/1	1.20
	#2/0 ACSR – 6/1	3.15
	#2 – 7 strand aluminum	.60
	#2/0 – 7 strand aluminum	1.15
	#4/0 – 19 strand aluminum	1.45
	336.4 MCM – 19 strand aluminum	1.50
	3/8" copperweld 7 – #8	5.60
8.4	Taps (cost each)	
	#2/0 run and tap, covered	3.85
	#4/0 run and tap, covered	4.85

TABLE 19. Continued

	Material and Equipment	Cost	
		1973	Current Price
	#2/0 aluminum, #2/9 ACSR, #2/0 copper	1.15	
	#4/0 aluminum, #4/0 ACSR, #4/0 copper	1.85	
	336.4 MCM ACSR, 300 MCM ACSR, 350 MCM copper	3.45	
8.5	Gutter taps, copper conductors (cost each)		
	#3/0 – 350 MCM run, #6 – #2 tap	3.35	
	#2/0 – 500 MCM run, 350 MCM tap	4.95	
	#3/0 – 500 MCM run, #1 – #4/0 tap	4.70	
	#3/0 – 500 MCM run, #4 – 500 MCM tap	5.75	
	400 MCM - 800 MCM run, #6 – #2 tap	8.05	
	400 MCM - 800 MCM run, #1 – #4/0 tap	8.65	
	400 MCM – 800 MCM run, 400 MCM 800 MCM tap	11.95	
	350 MCM – 800 MCM run, #2 – 800 MCM tap	5.85	
8.6	Solderless (cost each)		
	#6 solid copper	0.30	
	#4 solid copper	0.45	
	#2 stranded copper	0.50	
	#1 stranded copper	0.65	
	#2/0 stranded copper	0.85	
	#4/0 stranded copper	1.15	
8.7	Stirrups, wedge type (cost each)		

MATERIAL AND EQUIPMENT COSTS 53

	#2/0 copper or aluminum	2.90
	#4/0 copper or aluminum	3.00
	336.4 MCM copper or aluminum	9.95
8.8	Straight (cost each)	
	#1/0 to 300 MCM copper	4.35
	#1/0 to 300 MCM copper, heavy duty	17.25
	500 MCM to 800 MCM copper	11.00
	500 MXM to 800 MCM copper, heavy duty	30.50
8.9	Tee, copper conductor (cost each)	
	#1/0 – 300 MCM run, 300 MCM tap	11.50
	300 MCM – 500 MCM run, 300 MCM tap	12.15
	500 MCM – 800 MCM run, 800 MCM tap	17.35
8.10	Underground (cost each)	
	Aluminum connector, tapered #2	1.95
	Aluminum connector, tapered #1	2.45
	Aluminum connector, tapered #2/0	3.10
	Aluminum connector, tapered #4/0	3.85
	Aluminum connector, tapered 250 MCM	5.60
	Aluminum connector, tapered 500 MCM	8.40
9	CONTROL ITEMS (cost each)	
	115/230-V battery charger: 12-A switchboard, 30"x24"x7" panels (control boards for breakers), with average drilling, mounting of relays, switches, wiring, etc.	1,380. 6,325.

TABLE 19. Continued

Material and Equipment	Cost 1973	Current Price
10 COUPLING CAPACITOR POTENTIAL DEVICE (cost each)		
230 kV	2,201.	
115 kV	1,394.	
69 kV	1,095.	
46 kV	975.	
34.5 kV	915.	
Wave trap		
800 A	1,631.	
400 A	1,087.	
11 CROSSARMS (cost each)		
4" x 5" x 8'0" single	12.30	
4" x 5" x 8'0" double	25.00	
4" x 5" x 10'6" single	15.20	
4" x 5" x 10'6" double	31.05	
5" x 6" x 8'0" single	16.40	
5" x 6" x 8'0" double	36.90	
5" x 6" x 10'6" single	20.40	
5" x 6" x 10'6" double	43.55	
5" x 6" x 22'0" single	32.45	
5" x 6" x 22'0" double	65.00	

MATERIAL AND EQUIPMENT COSTS 55

	6" x 8" x 25'0" single	49.45
	6" x 8" x 25'0" double	96.60
12	CURRENT TRANSFORMERS (cost each)	
	5 kV – 25/5 A	149.50
	5 kV – 50/5 A	149.50
	5 kV – 100/5 A	149.50
	5 kV – 200/5 A	161.00
	15 kV – 15/5	277.15
	15 kV – 100/5 A	277.15
13	CUTOUTS (cost each)	
	7.8 kV – 100 A, open type	21.95
	7.8 kV – 200 A, open type	36.00
	15 kV – 100 A	24.95
	27 kV – 100 A	37.20
	Disconnect blades	
	7.8 kV – 300 A	31.05
	15 kV – 200 A	36.90
14	DEAD-END GRIPS (preformed) (cost each)	
	#2 aluminum, 15 kV	2.30
	#2/0 aluminum, 15 kV	3.00
	#4/0 aluminum, 15 kV	3.45
	#4/0 ACSR, 15 kV	3.60
	#4/0 ACSR, 23 kV	4.30
	#4/0 ACSR, 34.5 kV	5.75

TABLE 19. Continued

Material and Equipment	Cost 1973	Current Price
336.4 MCM aluminum, 15 kV	4.45	
336.4 MCM aluminum, 23 kV	6.80	
336.4 aluminum, 34.5 kV	7.15	
15 GROUNDING (cost each)		
Ground plate, 3' x 5'	51.20	
Ground plate platform	71.30	
Ground rod (copper covered)	3.95	
Ground rod (sectional)	3.75	
16 GUYS (cost each)		
Anchors (cost each)		
7,000-lb anchor plate	4.35	
12,000-lb anchor plate	7.15	
19,000-lb anchor plate	13.70	
Anchor rod, twin eye, ¾" x 8'0"	7.15	
Anchor rod, triple eye, 1" x 8'0"	14.40	
7,000-lb aluminum-coated steel guys (above 12 kV)	15.95	
12,000-lb aluminum-coated steel guys (above 12 kV)	16.40	
19,000-lb aluminum-coated steel guys (above 12 kV)	35.65	

Guy grips (cost each)
    3/8" Copperweld strand   2.30
    ½" Copperweld strand   3.90
    3 – #8 Alumoweld   0.80
    7 – #7 Alumoweld   2.40
    7 – #9 Alumoweld   1.15

17 INSULATORS AND ACCESSORIES

Glass strain, guy (cost each)
    Voltages up to 46 kV   11.95
    115 kV   14.95
    230 kV   19.55

Line post (cost each)
    12.5 kV, horizontal mounting, clamp type   13.80
    15 kV, switch and bus   12.65
    34.5 kV, switch and bus   23.60
    34.5 kV, vertical, clamp type   11.85
    34.5 kV, horizontal mounting, clamp type   14.95
    46 kV, vertical, clamp type   19.65
    46 kV, switch and bus   34.75

Suspension (cost each)
    7½" – 2.4 to 12.47 kV general use   3.75
    10" – 15,000-lb tension   4.60
    10" – 25,000-lb tension   5.60

Accessories (cost each)
    Anchor shackle   1.15
    Bell clevis   1.50

TABLE 19. Continued

Material and Equipment	Cost	
	1973	Current Price
Socket eye	1.15	
Pole gain, 12.5-kV construction	2.55	
18 LIGHTING		
Brackets (cost each)		
2' steel, 2" for fluorescent	59.75	
4' steel, 2" for fluorescent	75.90	
6' steel, 2" for fluorescent	92.30	
4' aluminum (1¼" wide)	6.00	
6' steel (2" wide)	12.20	
6' aluminum (1¼" wide)	11.50	
8' steel (2" wide)	22.55	
8' aluminum (1¼" wide)	19.00	
10' steel,(2" wide)	21.40	
12' steel (1¼" wide)	19.45	
16' steel (2" wide with 1¼" slipfitter end)	42.00	
20' steel (2" wide with 1¼" slipfitter end)	63.50	
Mercury vapor lamps, clear (cost each)		

Watts	Lumens		
100	3,650	5.75	
175	7,300	9.30	
250	11,000	9.80	
400	20,000	10.45	

MATERIAL AND EQUIPMENT COSTS

700	37,000	15.75
1,000	56,000	18.70

Mercury vapor lamps, color-corrected (cost each)

Watts	Lumens	
100	3,650	7.50
175	7,075	7.70
250	10,575	10.05
400	19,250	10.65
700	33,900	18.40
1,000	51,500	20.05

19 LINE GUARDS (cost each)

#4 ACSR, 7/1 single arm	0.70
#4 ACSR, 7/1 double arm	0.90
#2/0 ACSR, 6/1 single arm	1.15
#2/0 ACSR, 6/1 double arm	1.50
#4/0 ACSR, 6/1 single arm	1.25
#4/0 ACSR, 6/1 double arm	1.75
226.8 MCM ACSR, 6/7 single arm	1.60
226.8 MCM ACSR, 6/7 double arm	2.20
336.4 MCM ACSR, 26/7 single arm	1.75
336.4 MCM ACSR, 26/7 double arm	2.30
477 MCM ACSR, 24/7 single arm	2.60
477 MCM ACSR, 24/7 double arm	2.65

20 POLES (cost each)

Cedar (class 1)

40' long	78.00
45'	97.00
50'	120.00

TABLE 19. Continued

Material and Equipment	Cost	
	1973	Current Price
55'	152.00	
60'	207.00	
65'	299.00	
70'	336.00	
75'	403.00	
80'	472.00	
Cedar (class 2)		
40' long	74.00	
45'	92.00	
50'	106.00	
55'	147.00	
60'	196.00	
65'	276.00	
70'	311.00	
75'	368.00	
80'	431.00	
Yellow pine (class 3)		
40' long	71.00	
45'	81.00	
50'	95.00	
Yellow pine (class 4)		
40' long	62.00	

MATERIAL AND EQUIPMENT COSTS 61

	45'	64.00
	50'	84.00
21	POTENTIAL TRANSFORMERS (cost each)	
	24 kV, 200/1 turns ratio	1,795.00
	34.5 kV, 300/1 turns ratio	2,245.00
	46 kV, 400/1 turns ratio	2,300.00
	230 kV, 100 kVA	38,030.00
	230kV, 50 kVA	22,600.00
	115 kV, 100 kVA	14,895.00
	115 kV, 50 kVA	11,085.00
	69 kV. 4.5 kVA	3,325.00
	46 kV, 4.5 kVA	2,290.00
	34.5 kV, 3 kVA	1,945.00
	12.4 kV, 3 kVA	1,640.00
22	SWITCHES	
	3-pole switches (cost each)	
	230 kV, double break, gang operated, 1,600 A	9,145.00
	230 kV, double break, gang operated, 1,200 A	8,640.00
	230 kV, vertical break, gang operated, 1,600 A	8,330.00
	230 kV, vertical break, gang operated, 1,200 A	7,195.00
	115 kV, double break, gang operated, 1,200 A	4,645.00
	115 kV, double break, gang operated, 600 A	3,610.00
	115 kV, vertical break, gang operated, 1,200 A	4,065.00
	115-kV, vertical break, gang operated, 600 A	3,450.00
	Operating mechanism for switches (cost each)	485.00
	69 kV, gang operated, 1,200 A	3,220.00

TABLE 19. Continued

Material and Equipment	Cost	
	1973	Current Price
69 kV, gang operated, 600 A	2,415.00	
46 kV, gang operated, 1,200 A	2,300.00	
46 kV, gang operated, 600 A	1,725.00	
34.5 kV, gang operated, 1,200 A	1,670.00	
34.5 kV, gang operated, 600 A	1,265.00	
23 kV, gang operated, 1,200 A	1,495.00	
23 kV, gang operated, 600 A	980.00	
14.4 kV, gang operated, 1,200 A	1,265.00	
14.4 kV, gang operated, 600 A	825.00	
34.5 kV, 600 A, single arm, horizontal mounting	825.00	
15 to 23 kV, 600 A, vertical mounting	725.00	
15 to 23 kV, 600 A, single arm, horizontal mounting	690.00	
15 to 23 kV, 400 A, single arm, horizontal mounting	575.00	
Air break switches (cost each)		
46 kV, 600 A (with handle and wood insulator)	1,495.00	
34.5 kV, 600 A (with handle and wood insulator)	1,495.00	
34.5 kV, 400 A (with handle and wood insulator)	1,485.00	
14 kV, 400 A (with handle)	655.00	

MATERIAL AND EQUIPMENT COSTS 63

1-pole switches (cost each)	
115 kV, 1,200 A	715.00
115 kV, 600 A	600.00
69 kV, 1,200 A	415.00
69 kV, 600 A	310.00
46 kV, 1,200 A	265.00
46 kV, 600 A	185.00
34.5 kV, 1,200 A	220.00
34.5 kV, 600 A	150.00
23 kV, 1,200 A	175.00
23 kV, 600 A	105.00
14.4 kV, 1,200 A	160.00
14.4 kV, 600 A	100.00
23 TERMINALS (cost each)	
15 kV cable	
#4/0 Aluminum conductor, 400 A	94.00
500 MCM aluminum conductor, 600 A	95.00
25 kV cable	
#4/0 aluminum conductor, 400 A	106.00
500 MCM aluminum conductor, 600 A	106.00
35 kV cable	
#4/0 aluminum conductor, 400 A	178.00
500 MCM aluminum conductor, 600 A	178.00
1 15 kV lead break terminators	32.00
23 kV non lead break terminators	41.00

TABLE 19. Continued

	Material and Equipment	Cost	
		1973	Current Price
24	TRANSFORMERS (cost each)		
	Padmount (single phase) 4.16 GY/2.4 kV		
	15 kVA	300.00	
	25 kVA	390.00	
	50 kVA	625.00	
	75 kVA	900.00	
	100 kVA	1,095.00	
	167 kVA	1,655.00	
	Padmount (single phase) 34.4 GY/20 kV		
	15 kVA	795.00	
	25 kVA	865.00	
	50 kVA	1,140.00	
	100 kVA	1,585.00	
	Padmount (three phase) 4160 GY/2400 V 208Y/120 V		
	75 kVA	2,505.00	
	112.5 kVA	2,680.00	
	150 kVA	2,900.00	
	225 kVA	3,335.00	

300 kVA	3,690.00
500 kVA	4,715.00
750 kVA	6,070.00
1,000 kVA	8,225.00
1,500 kVA	10,455.00
2,000 kVA	12,220.00

Submersible (single phase) 4160/12470 V

15 kVA	460.00
25 kVA	535.00
50 kVA	780.00
100 kVA	1,100.00
167 kVA	1,575.00

Submersible (single phase) 24900/14400 V

15 kVA	485.00
25 kVA	560.00
50 kVA	810.00
100 kVA	1,135.00
167 kVA	1,575.00

Submersible (three phase) 4160 GY/2400 V / 208Y/120 V

45 kVA	1,610.00
75 kVA	2,020.00
112.5 kVA	2,820.00
150 kVA	3,185.00
225 kVA	4,695.00

TABLE 19. Continued

Material and Equipment	Cost	
	1973	Current Price
300 kVA	5,805.00	
500 kVA	8,225.00	
Submersible (three phase) 22860 GY/13200 V 208Y/120 V		
45 kVA	1,775.00	
75 kVA	2,220.00	
112.5 kVA	3,105.00	
150 kVA	3,500.00	
225 kVA	5,165.00	
300 kVA	6,105.00	
500 kVA	9,050.00	

APPENDICES

APPENDIX A. ABBREVIATIONS USED IN TEXT AND TABLES

A	ampere
AAC	all aluminum conductor
ac	alternating current
ACSR	aluminum conductor, steel reinforced
AWA	Alumoweld aluminum
CCPD	coupling capacitor potential device
CM	circular mil
CT	current transformer
CWD	Copperweld (wire)
dc	direct current
EHS	extra high strength
GY	grounded wye system
HDBC	hard drawn bare copper
IPS	iron pipe size
kV	kilovolt
kVA	kilovolt-ampere
kVAR	kilovar (reactive kilovolt-ampere)
M	unit of 1000
MCM	circular mil × 1000
OCB	oil circuit breaker
PT, POT TRANS	potential transformer
PVC	polyvinylchloride
SPST	single pole, single throw
STR	strand
V	volt
w/cond	with conduit
7C No. 9	7 conductor, No. 9 wire
12C No. 10	12 conductor, No. 10 wire
26/7	26 strand, No. 7 wire

APPENDIX B. GLOSSARY

armor rod	A spiral layer of round rods surrounding the conductor for a short distance. These rods reduce vibration and protect against chafing and flashover at point of support. Used on high-voltage transmission lines.
battery charger	Device used to charge a storage battery. When operating on alternating current, a rectifier is used.
breakdown voltage	The voltage at which the dielectric strength of an insulator breaks down and arcing begins.
bus or bus bar	The main circuit to which generators and feeders are connected in a power station. A heavy solid wire that connects points at the same potential.
bus (auxiliary)	A second bus that may have a different voltage from the main bus.
bus support stand	Support for buses, disconnect switches, and auxiliary equipment.
bushing	An insulating tube that protects a conductor as it passes through a hole in a support structure or apparatus.
cable	A stranded conductor or several insulated conductors used for power transmission or distribution.
cable (coaxial)	A two-conductor transmission line in which the inner conductor is separated

cable rack	from the outer, tubular conductor by a dielectric material. A frame for supporting electrical cables.
capacity	Maximum power output of a motor (horsepower) or of a generating station (megawatts). Also, the energy-carrying rating of an electrical device, wire, or cable.
carrier current	A high-frequency current superimposed on the normal frequency of a power transmission line for communication or telemetering control.
center of distribution	A point near the center of the area served by a feeder from a power station or substation.
circuit breaker	A device used to open a current-carrying circuit automatically under abnormal conditions without damage to itself.
circuit breaker (oil)	A high-voltage circuit breaker with contacts immersed in oil. The oil acts as a coolant and extinguishes the arc.
circular mil	A unit used to designate the cross-sectional area of a round conductor. One circular mil is the area of a circle one mil in diameter.
cluster	A lighting fixture having two or more lamps.
coupling capacitor	The coupling between circuits or components by means of a capacitor.
cutout	A device that disconnects one circuit from another.
dead end	The point in a transmission line where all the strain in the conductors is carried by a support.
dead-end tower	A structure that withstands the strain of dead-ending long-line spans.

disconnect	To remove an electrical device from a circuit or to unfasten a wire.
disconnect switch	A switch for cutting out high-voltage circuits.
distribution	The supply of power to points of utilization from the source of generation.
frequency	The number of cycles per unit time in alternating current.
generator	A rotary machine for converting mechanical energy to electrical energy.
ground	A common return to a point whose potential is taken as zero, i.e., the earth.
ground switch	A switch used to ground deenergized high-voltage lines before work of any kind is done on these circuits.
guy	One or more braces or cables used to stiffen a pole and keep it in position.
insulator	Device used to isolate electrical conductors from their grounded supports.
insulator (post-type)	A porcelain insulator that is made in one piece for low-voltage lines, and of two, three, or four layers cemented together to form a rigid unit for high-voltage applications.
insulator (strain)	An insulator that must also withstand mechanical stress. Used where a line is dead-ended, at sharp curves, and at extra long spans. Also used to insulate lower part of guy cable from ground.
insulator (suspension)	An insulator that is suspended from a crossarm or structure and supports the line conductor. The insulators can be placed in a string, the number of units depending on the line voltage.
junction box	Box where circuit is connected to a main.

APPENDIX B. GLOSSARY 73

lightning arrester	A device which protects equipment from the destructive effects of lightning surges. It allows the high-voltage surge to pass directly to ground without allowing ordinary current to follow.
lightning mast	A steel mast or extension of a steel structure that diverts all lightning which might otherwise strike a bus, disconnecting switch, etc.
line guard	Same as armor rod, except smaller in size. Used on lower voltage conductors.
lineman	A man who erects or works on electric transmission lines or structures.
main structure	Structure where terminations are made from incoming or outgoing transmission lines. Supports high-voltage buses, disconnecting switches, lightning arresters, etc.
potential	The voltage or electrical pressure that exists between two points in a circuit.
pothead	A container with a hole in its bottom and a flared top. The lead sheath of a cable is inserted through the hole and the pot is filled with a molten insulating compound. Potheads prevent moisture from entering the insulation of the cable and serve to separate the conductors to prevent arcing between them. Used on cables terminating at buses or similar terminations above 750 V.
power fuse	A fuse used in high-capacity power circuits. In general, power fuses operate faster than circuit breakers at or near their interrupting ratings.
primary	The transformer's circuit (or winding) that is connected to the power source.

relay	A device installed in a system to electrically trip circuit breakers or contactors in order to isolate a short circuit or abnormal surge of power.
resistance	The opposition offered to the flow of an electrical current that results in the production of heat.
secondary	The transformer winding in which a voltage is induced from the primary winding.
switch	A device for closing, opening, or changing the connections of a circuit.
switchboard	A panel upon which the switches, rheostats, meters, relays, and control switch are mounted for the control of electrical machines, circuit breakers, and systems.
switch (double break)	A switch that connects and disconnects two contacts at the same time.
switch (horn gap)	A switch used to break relatively heavy currents, e.g., the sectionalizing of transmission lines.
switch (quick break)	A switch having spring action that breaks the circuit quickly when pulled open by hand.
terminal	A point to which electrical connections may be made.
tie-down stand	A tower used to secure the transmission line where it may be subject to uplifting action.
tower	Steel structure that supports cross-country transmission lines.
transformer	A device in which a varying current in the primary circuit produces, by electromagnetic induction, a voltage in the secondary.
transformer (current)	A transformer in which the required current in the secondary winding and the

APPENDIX B. GLOSSARY

	current in the primary determine the turn ratio. It is used with a meter when the line current is too great to be measured directly or when the circuit voltage is too high for safe operation of meters or other control equipment.
transformer (potential)	A transformer used to step the voltage down for voltmeters and other instruments.
transformer (power)	The transformer which supplies operating voltages to the various alternating-current circuits. The transformer obtains its power from the alternating-current line.
transformer (station service)	The transformer that provides electrical service to the control house, yard lighting, etc.
transformer (step-down)	A transformer in which the primary voltage is stepped down to a lower secondary voltage.
transformer (step-up)	A transformer in which the primary voltage is stepped up to a higher secondary voltage.
transmission line	High-voltage conductors that carry electric power from one place to another. The voltage is dependent upon the distance to be covered.
tuned circuit (line)	A circuit in which one or more components are adjustable to produce resonance at a desired frequency.
turning structure	A tower used for turning, or changing direction of the transmission line.
wave trap	A parallel resonant circuit tuned to offer a high impedance at a specific carrier frequency. It is inserted in series with one of the conductors of a transmission line.

APPENDIX C. SAMPLE BID PACKAGE FOR 115/69/34.5-KV SUBSTATIONS

This supplemental material is designed to familiarize a relatively inexperienced substation estimator with the procedures that accompany an estimate for bids. In addition, readers of all experience levels will benefit by preparing a sample estimate from the following data and comparing results with those of the author. A summary sheet of the author's estimate appears as Table 21. The form shown as Table 18 is recommended for use.

A typical bid package contains a bid invitation, specifications, a bill of material, and drawings. A single-line diagram (Fig. 14) is always included and indicates clearly whether circuits are overhead or underground, and if they are for present or future use. Power, station service, and instrument transformers are also shown, along with their ratings, as are fuses, oil circuit breakers, and lightning arresters. The remaining drawings (Figs. 15, 16, and 17) are plan, elevation, and sectional views showing the physical layout and dimensions of the structures and equipment.

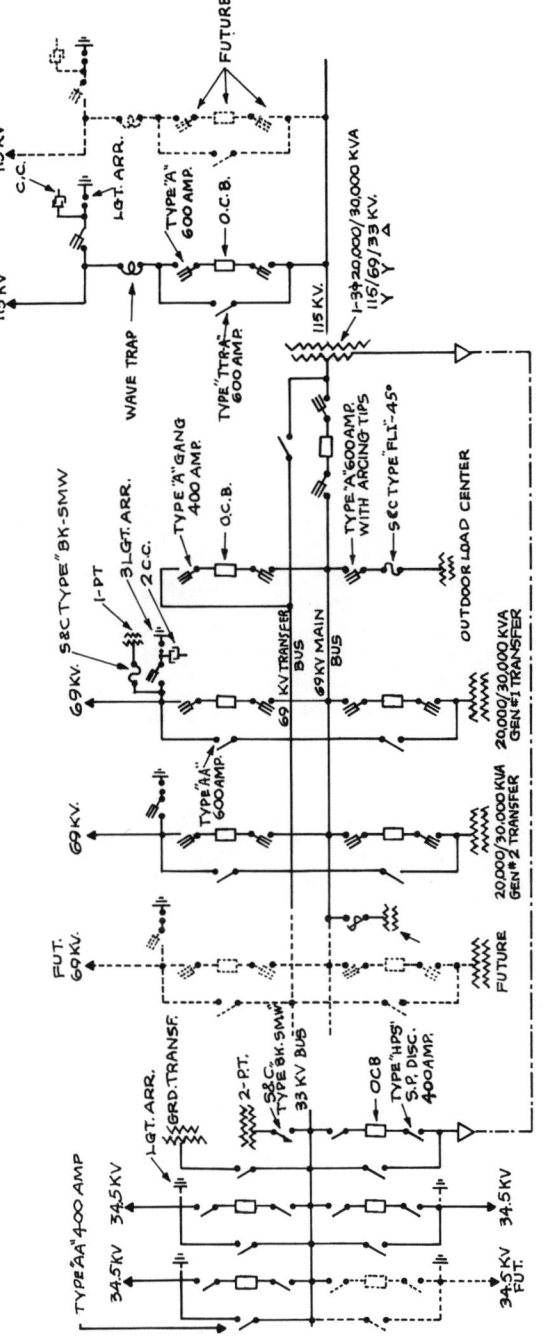

Figure 14. Single-line diagram for a 115/69/34.5-kV substation.

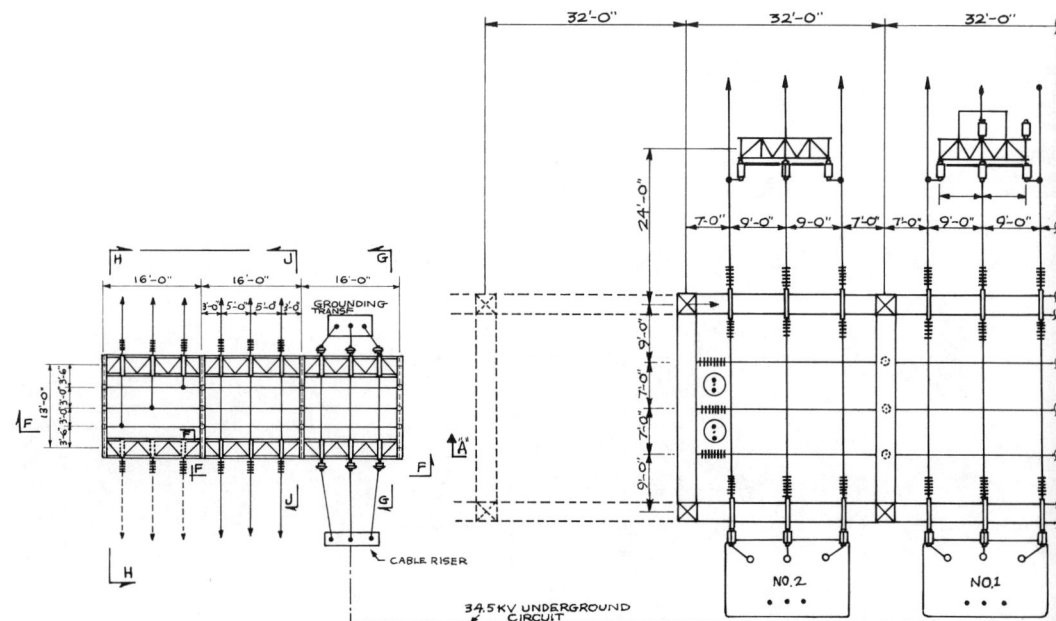

Figure 15. Plan view of a 115/69/34.5-kV substation.

APPENDIX C. SAMPLE BID PACKAGE FOR 115/69/34.5-kV SUBSTATION 79

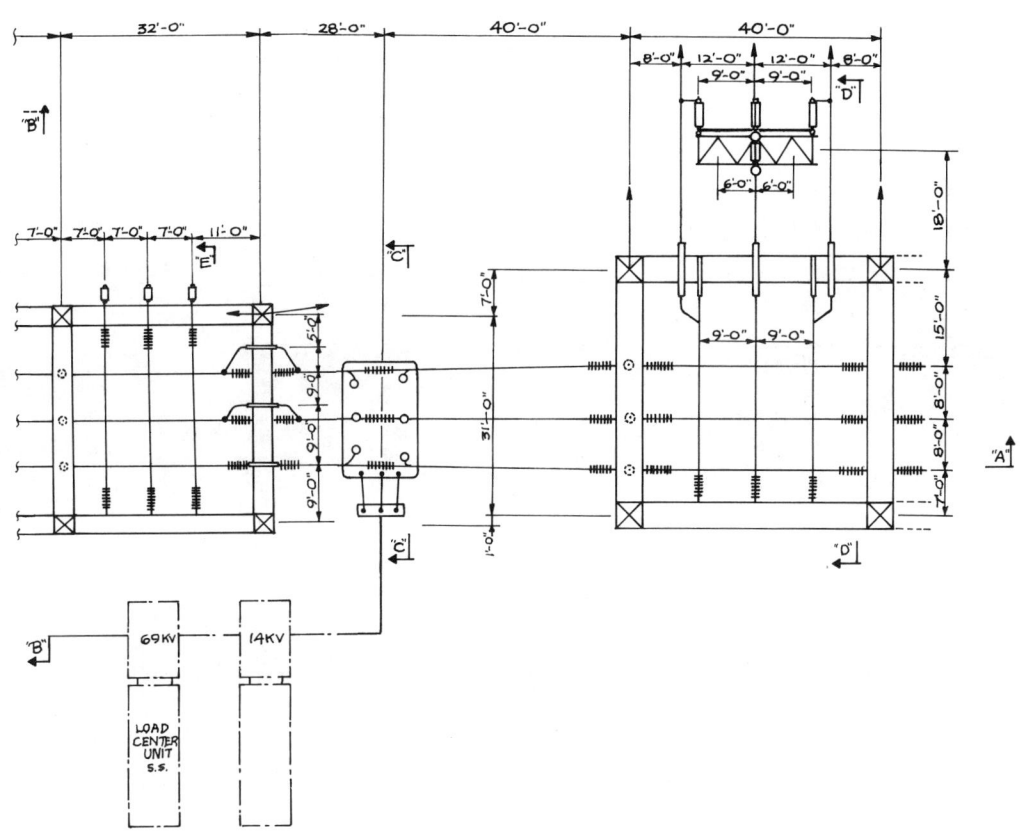

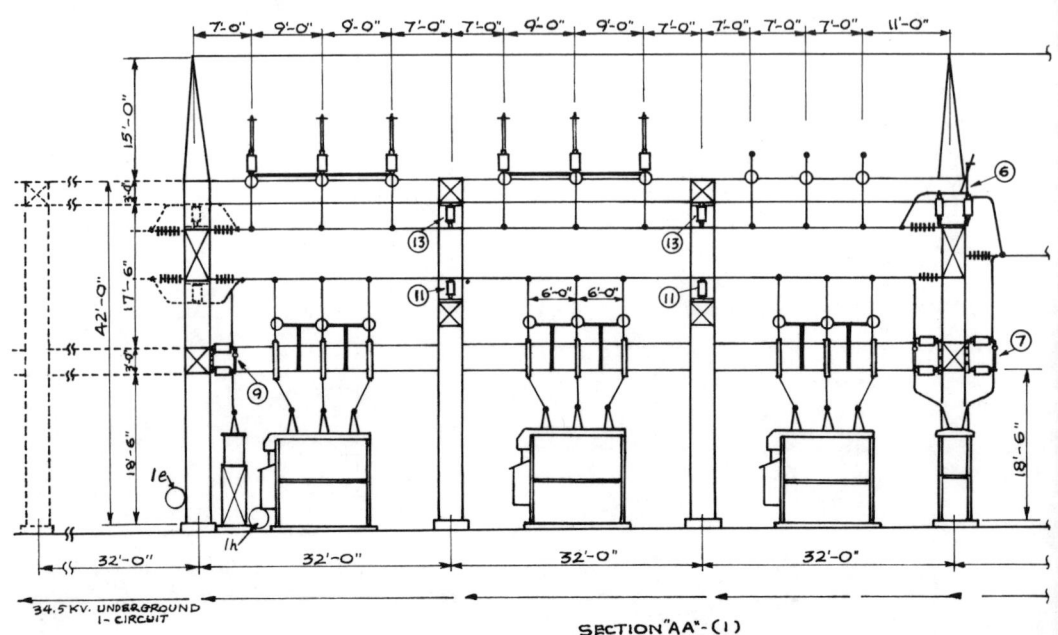
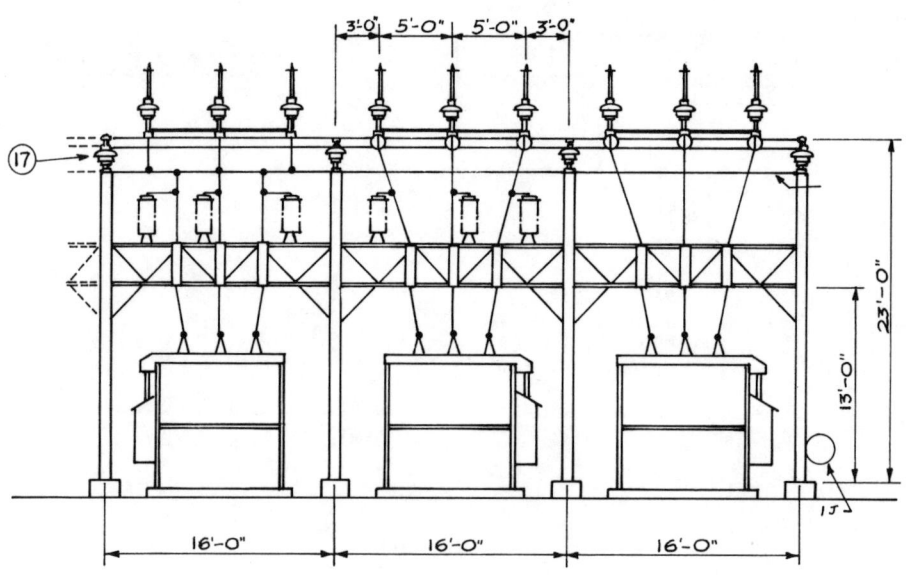

Figure 16. Sectional views AA, FF, GG, HH, and JJ of a 115/69/34.5-kV substation.

APPENDIX C. SAMPLE BID PACKAGE FOR 115/69/34.5-kV SUBSTATION 81

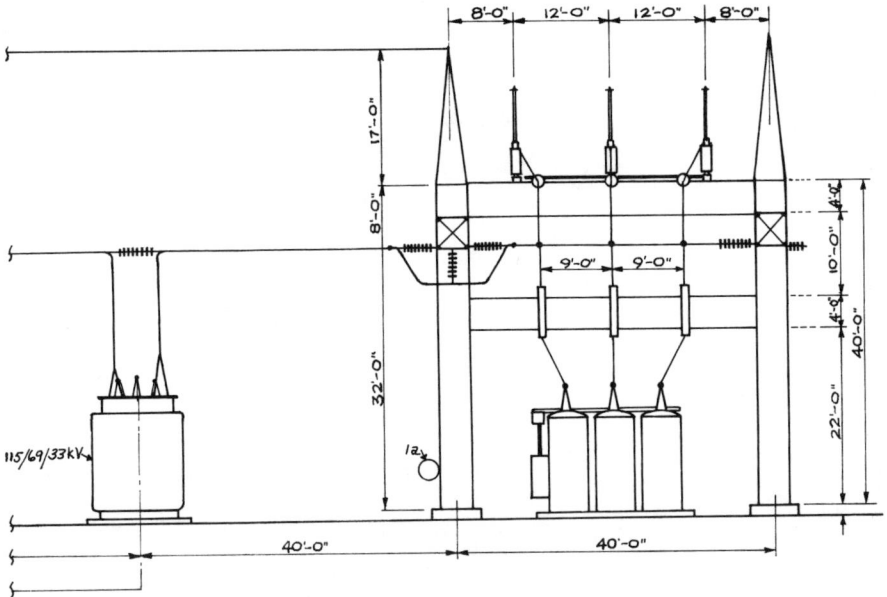

SECTION "AA"-(2)

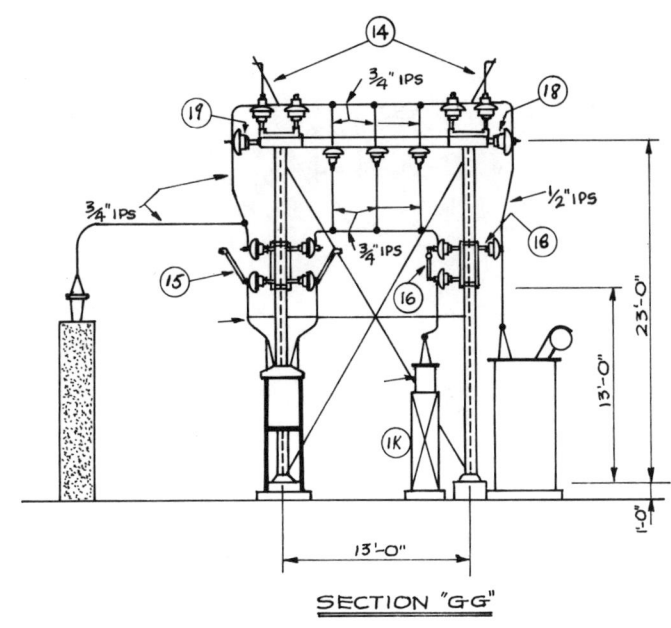

SECTION "GG"

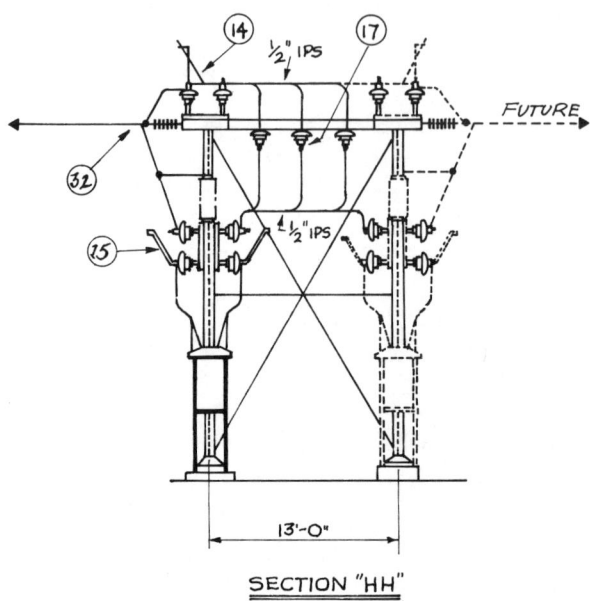

SECTION "HH"

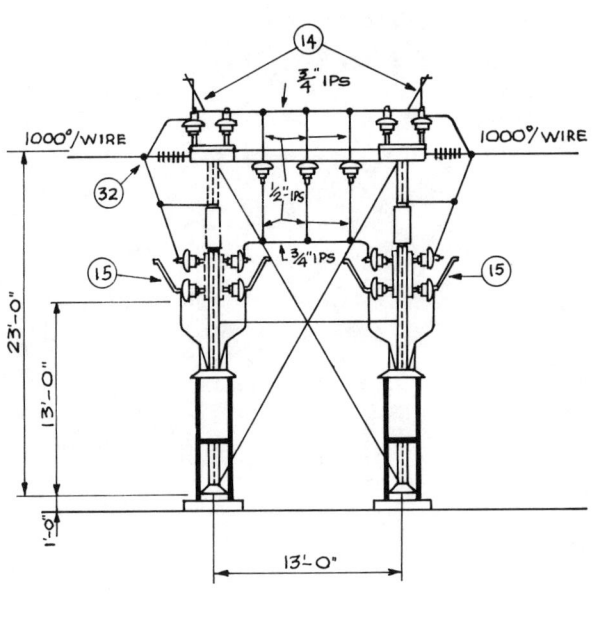

SECTION "JJ"

APPENDIX C. SAMPLE BID PACKAGE FOR 115/69/34.5-kV SUBSTATION   83

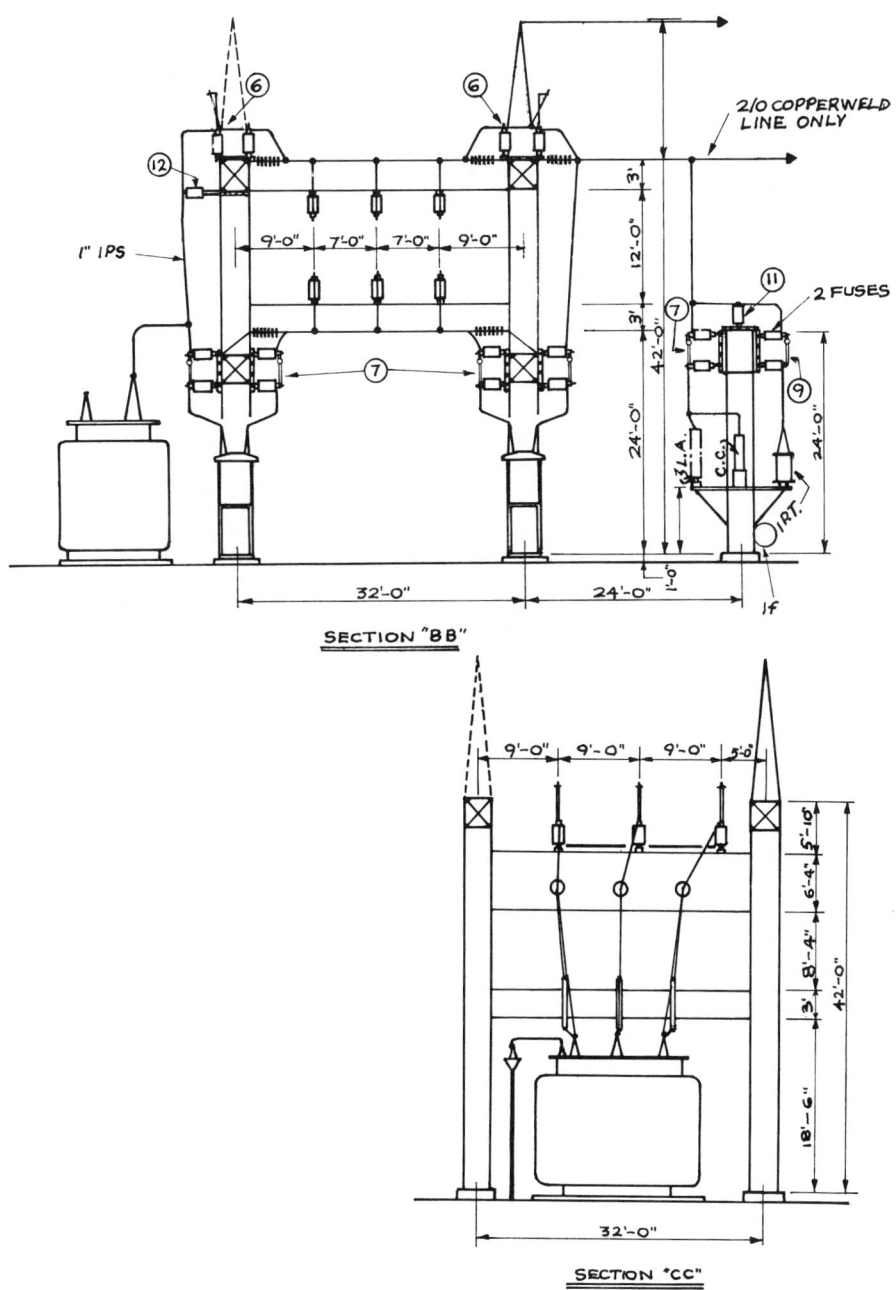

Figure 17. Sectional views BB, CC, DD, and EE of a 115/69/34.5-kV substation.

84 ELECTRICAL POWER SUBSTATIONS

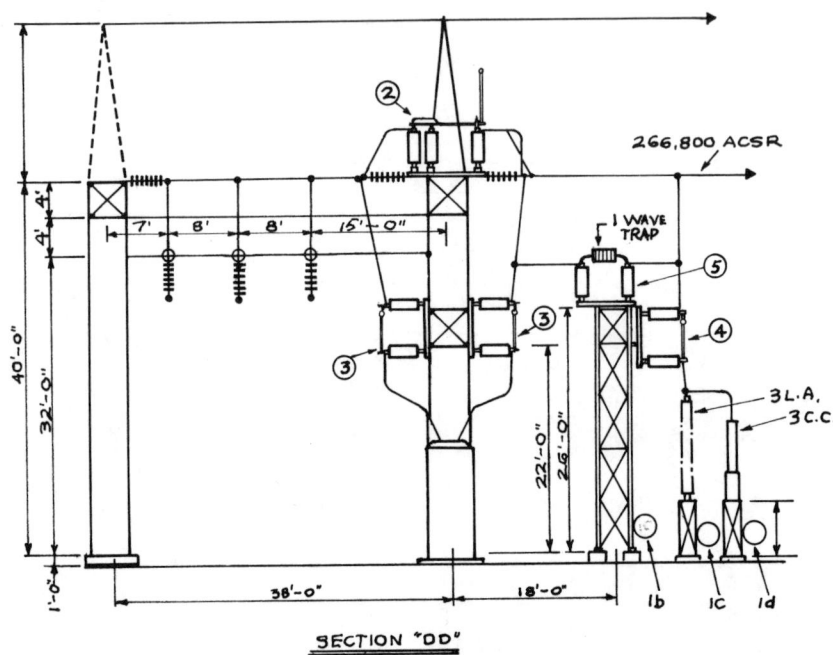

SECTION "DD"

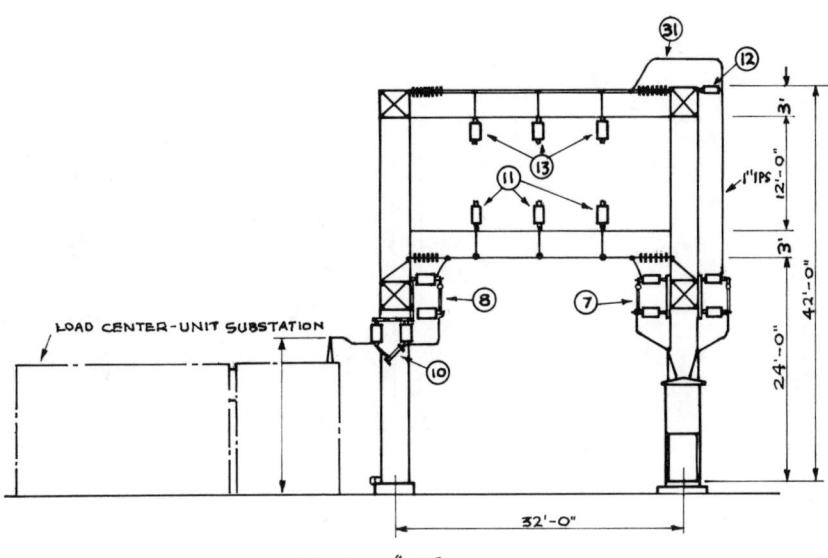

SECTION "EE"

APPENDIX C. SAMPLE BID PACKAGE FOR 115/69/34.5-kV SUBSTATION

Invitation To Bid on Construction of Substation

The bid invitation usually includes a cover letter that explicitly sets forth the bidding requirements and the qualifications of the bidders. The bidding instructions outlined below are self-explanatory. Usually from five to seven companies compete for a job of this type. The contractor is usually given 2 weeks to prepare his bid, but may be asked to submit one in as little as 1 week. Of course, the lowest responsible bidder can be expected to obtain the job.

The following is a sample cover letter.

General. Sealed proposals, addressed to the purchaser, will be received no later than 11:00 a.m. on Tuesday, October xx, 19xx, at the office of the purchaser located at (address), at which time they will be (publicly) (privately) opened.

Required Documents. All documents in the bid package shall be returned with the bid. Each bidder is cautioned to study all parts of the specification and bill of material, for he shall be held responsible for costs of each item as well as his total bid. The proposal must be signed by a legally responsible member of the bidder's organization.

Bid Security. Security in the amount of at least 10% of the total bid shall be required from each bidder. The bid bond shall be a guarantee to the purchaser that if he accepts the proposal, the contract will be executed and the performance of the contract properly secured. Acceptance of a bid bond shall not be construed to limit the contractor's liability for failure to execute the contract. If a bid is not accepted, the security shall be returned after the contract has been awarded or within 30 days after receipt of the bids.

Bonds and Insurance. The successful bidder must submit evidence satisfactory to the purchaser, that the required bonds and insurance have been obtained. Such evidence must be presented within 5 days after the award of the contract.

Withdrawal or Rejection of Bids. A bidder may withdraw his proposal any time before the official receiving date. No bid may be withdrawn for 30 days after the official date of receipt. The purchaser may reject any or all bids and may waive any defect or informality in any bid.

Responsible Bidders. Only responsible contractors will be authorized to do the work. Bidders shall be familiar with the type of construction to be used and the labor conditions involved. The purchaser reserves the right to reject a bidder because of inadequate labor, equipment or experience, and may require the bidder to provide evidence that each of these requirements can be met.

Schedule of Work. After receipt of bids, the purchaser may request that the contractor prepare a construction schedule and submit it to the purchaser for approval. It will be used by the purchaser for assurance that the work will be completed in the specified time. This schedule shall be followed by the successful bidder unless specific changes are approved by the purchaser.

General Specifications for Substation

Scope of Work. The contractor must furnish all labor, or labor and materials, tools, equipment, transportation, superintendence, and services required to complete the job as listed on the bill of material, line diagram, and plan and sectional drawings.

All prints are to be returned with the bids. The company reserves the right to reject any and all bids.

Records. The contractor is to keep separate construction and maintenance cost records. Maintenance costs include moving existing equipment and transformers, keeping circuits in operation temporarily, etc.

The contractor will submit a record of all equipment installed and removed.

The contractor is to submit daily time sheets specifying time spent on either construction or maintenance.

APPENDIX C. SAMPLE BID PACKAGE FOR 115/69/34.5-kV SUBSTATION

All invoices for work performed should be approved by the field project engineer before being submitted for payment. Invoices for extra work should be clearly marked or stamped *extra work*.

Extra Work. Any additional work required of the contractor that is not covered will be performed on the basis of a cost plus percentage (percentage will represent contractor's profit only). Contractor should specify percent amount in bid. No additional work is to be performed without written authorization of the company construction engineer or his representative. This authorization must be obtained before additional work begins.

Material. The contractor shall furnish labor and all remaining material necessary to complete work. The company will provide materials except where noted.

Tools and Equipment. All tools and equipment used in normal performance of work will be supplied by the contractor.

The contractor will include the cost of vehicles and special equipment in the bid. This will include expendable items such as drill bits, hacksaw blades, canvas, brooms, welding rods, and gas. A schedule of rates charged for all vehicles and equipment that are expected to be used should be included in bid along with their exact description.

Written authorization must be obtained before use of all vehicles and equipment. Request for authorization should be made to the project engineer.

Labor Rates. A schedule of rates, submitted with this bid, shall include the following information:
- Hourly rate and classification of all men employed. Include a schedule of built-up rates* if applicable to your company. Include also regular, overtime, and emergency rates.
- Rate for superintendent and general line foreman.
- Rate for clerical employees in the home office.

*Built-up rates are the base hourly rates plus percentage costs for insurance, Social Security, administrative overhead, etc. An example of this is shown on page 18.

88 ELECTRICAL POWER SUBSTATIONS

Safety Rules. The following safety rules will be observed:

- Extension cords and electric tools. Motor frames of portable electric tools shall be properly grounded when connected to energized circuits.
- Safety hats. Safety hats will be furnished by the contractor to all his personnel. They shall be worn at all times while working on substation property. They shall be worn in an area where others are working above them or where materials or equipment may fall during the course of the work.
- Safety belts. When working on poles, steel structures, or other elevated supports, employees shall wear a safety belt, strap, or sling whenever possible.
- Ladders. Ladders must be placed at a safe working angle and be fitted with shoes or spurs to prevent slipping. Where the ground surface is slippery or where other hazardous conditions exist, another workman shall hold the ladder steady or it shall be otherwise secured. (It is good practice to rope lash a ladder in position whenever possible.) Portable metal ladders or those with metal siderails or rungs shall not be used where there is danger of contact with energized parts.
- Housekeeping. Good housekeeping at the job site is fundamental to continuity of service and essential for the prevention of accidents and fires.

Insurance. The successful bidder must file with the company a certificate of insurance covering Workmen's Compensation, bodily injury, and property damage. This coverage is to be furnished at no cost to the company. Further, the successful bidder shall provide sufficient cargo insurance, at no cost to the company, if called upon to transport company equipment that is related to the job.

Inspection and Tests. The company has rights of inspection for compliance to the terms of this specification. Tests are to be performed by the company unless otherwise noted.

APPENDIX C. SAMPLE BID PACKAGE FOR 115/69/34.5-kV SUBSTATION

Bill of Material for Substation

It should be possible to find the man-hour requirements for most of the foregoing items in Tables 1 through 6. Because of the enormous task of acquiring data for every component, there will be gaps that the reader must work around until he develops his own information. For example, this book contains practically no man-hour data on 69-kV components. It is suggested that time values for similar components with higher and lower voltage ratings be found in the tables and that the reader choose an intermediate value as an approximate estimate of the labor involved.

The author has added notes (italic) to the following sample bill of material (Table 20) to show his assumptions and estimates of man-hours for the unlisted components.

TABLE 20. Bill of Material for 115/69/34.5-kV Substation
(Material Furnished by Company Unless Otherwise Stated)

Item No.	Quantity	Description
		STEEL DETAILS
1 a*	1	115-kV main structure (24,000 lb)
b	1	115-kV wave trap rack (200 lb)
c	3	115-kV lightning arrester rack (200 lb each)
d	3	115-kV coupling capacitor potential device racks (300 lb each)
e	1	69-kV switching structure (19,000 lb)
f	1	69-kV lightning arrester and potential transformer rack (400 lb)
g	1	69-kV lightning arrester rack (200 lb)
h	1	69-kV rack for two potential transformers (350 lb)
j	1	34.5-kV switching structure (15,000 lb)
k	1	34.5-kV rack for two potential transformers (300 lb)
		115-kV EQUIPMENT
2*	1	ITE, type TTR-A 600-A, 3-pole horn gap switch *(40 man-hours each)*
3*	2	Type X 600-A, 3-pole gang disconnect switches (vertical mounting) *(40 man-hours each)*

*Illustrated on drawings which start on page 77.

TABLE 20. Continued

Item No.	Quantity	Description
4*	1	Type A 600-A, 3-pole gang disconnect switch (vertical mounting) *(40 man-hours each)*
5*	1	Wave trap rack mounting consisting of 2 insulator stacks (3 units per stack), two 5" to 3" adapters and one channel base (400 lb)

69-kV EQUIPMENT

Item No.	Quantity	Description
6*	5	Type AA 600-A, 3-pole horn gap switch and accessories *(35 man-hours each)*
7*	14	Type A 600-A, 3-pole gang disconnect switches (vertical mounting) *(35 man-hours each)*
8*	1	Type A 600-A, 3-pole gang disconnect switch with arcing tips (vertical mounting) *(35 man-hours each)*
9*	5	S & C Fused disconnect switch *(2 man-hours each)*
10*	3	S & C fuse holders with clip *(0.5 man-hours each)*
11*	3	Bus support—69kV (2/0 strand) *(2 man-hours each)*
12*	9	Bus support—69kV (1" IPS) *(2 man-hours each)*
13*	6	Bus support—69 kV(1¼" IPS) *(2 man-hours each)*

34.5-kV EQUIPMENT

Item No.	Quantity	Description
14*	5	Type AA 400-A, 3-pole horn gap switches *(30 man-hours each)* Note: Switches have terminals as follows: Two 3-pole switches, jaw for 2/0 strand, hinge for ¾" IPS One 3-pole switch, jaw for ½" IPS, hinge for ¾" IPS One 3-pole switch, jaw for 2/0 strand, hinge for ½" IPS One 3-pole switch, jaw and hinge for ¾" IPS
15*	24	Type HPS 400-A, single pole disconnect switch (vertical mounting) *(use man-hours for type HPL listed in Section 2)* Note: Switches have terminals as follows: Nine 1-pole switches, jaw and hinge for 2/0 strand

*Illustrated on drawings which start on page 77.

APPENDIX C. SAMPLE BID PACKAGE FOR 115/69/34.5-kV SUBSTATION

TABLE 20. Continued

Item No.	Quantity	Description
		Six 1-pole switches, jaw for ¾" IPS, hinge for 500 MCM
		Six 1-pole switches, jaw for ¾" IPS, hinge for 2/0 strand
		Three 1-pole switches, jaw for ½" IPS, hinge for 2/0 strand
16*	3	S & C fused disconnect switch *(2 man-hours each)*
		Terminals on switches:
		top for ¾" IPS
		bottom for 2/0 strand
17*	12	Bus support—34.5 kV (1¼" IPS) *(2 man-hours each)*
18*	6	Bus support —34.5 kV (½" IPS) *(2 man-hours each)*
19*	3	Bus support — 34.5 kV (¾" IPS) *(2 man-hours each)*
		BUS FITTINGS
20	6	bus tap, 1½" IPS run - ¾" tap
21	12	bus tap, 1¼" IPS run - ½" tap
22	6	bus tap, 1" IPS run - 2/0 strand tap
23	9	bus tap, ¾" IPS run and tap
24	6	bus tap, ¾" IPS run, ½" IPS tap
25	15	bus tap, 250 MCM run and tap
26	57	bus tap, 2/0 strand run and tap
27	10	Bus tap, Copperweld run, 2/0 strand tap
28	5	Bus tap, 266.8 MCM ACSR run, 250 MCM tap
29	6	Straight connector† for 1¼" IPS
30*	9	Straight connector† for 1" IPS
31*	3	End connector for 1" IPS to 2/0 strand
32	9	Parallel clamp† for 2/0 strand
33	24	115-kV string of strain insulators
34	57	69-kV string of strain insulators *(1.43 man-hours each)*
35	9	34.5-kV string of strain insulators
36	3	115-kV lightning arrester
37	6	69-kV lightning arrester *(2.80 man-hours each)*

* Illustrated on drawings which start on page 77.
†Bus taps, straight connectors, end connectors, and parallel clamps are same as "bus connections (not welded)" in Table 3.

TABLE 20. Continued

Item No.	Quantity	Description
38	9	34.5-kV lightning arrester
39	150'	1¼" IPS copper pipe
40	268'	1" IPS copper pipe
41	228'	¾" IPS copper pipe
42	162'	½" IPS copper pipe *(10 man-hours per 100')*
43	75'	500 MCM stranded copper cable
44	600'	250 MCM stranded copper cable *(11 man-hours per 100')*
45	2,500'	2/0 stranded copper cable *(11 man-hours per 100')*
		ELECTRIC EQUIPMENT
46	1	115-kV OCB (haul and place on foundation) *(70 man-hours)*
47	6	69-kV OCB (haul and place on foundation) *(30 man-hours each)*
48	4	34.5-kV OCB (haul and place on foundation) *(15 man-hours each)*
49	1	50-kVA station service transformer
		CONTROL HOUSE
50	11	One ac and one dc distribution panel, fixtures, conduit, wiring, etc. *(listed as 127 man-hours in Table 5)*
51	1	230-Volt battery charger and rack
52	1	Set of 20 6-Volt batteries and rack
53	11	Switchboards (control for breakers)
54	1	Supervisory control cabinet
55	1	Carrier control cabinet
56	2,400'	12C #10 cable
57	3,800'	4C #10 cable
58	1,200'	2C #10 cable
59	100'	#8 carrier control cable
60	160'	RG-34U coaxial cable
		YARD LIGHTING
61	8	Mercury light fixture (includes concrete pedestal, photocell, Luminaire mounting bracket)

TABLE 20. Continued

Item No.	Quantity	Description
62	6	Yard convenience outlet
63	6	3-phase oil filter outlet
64	300'	2" galvanized conduit
65	450'	1" galvanized conduit
66	630'	¾" galvanized conduit *(20 man-hours per 100')*
67	1,330'	2C #10 cable
68	1	1-pole, single throw tumbler switch
69		Junction boxes, Unilets, Sealtite, connectors, etc. to be included *(Assume 10 junction boxes to be installed, average 10" size)*

GROUNDING

70	4,700'	Install 4/0 HDBC ground grid, equipment ground, and manhole grounds (include Cadweld connections, ground connectors, trenching, and backfill)
71		Megger ground grid and record test results†

MISCELLANEOUS

72		Name plates and signs, breaker or circuit names *(assume 10 name plates to be installed)*†
73		Move equipment (to and from job site)†
74		Final clean-up†

EXTRA WORK

Any extra work will be on a cost plus 10% basis

List percentage cost for following on extra work:

                                      Percent of bare labor

    Insurance

    Social Security

    Administrative overhead

    Workmen's compensation

    Other (specify)

†Material supplied by contractor.

Author's Estimate

The author's estimate for the 115/69/34.5-kV substation is summarized in the table below.

TABLE 21. Author's Cost Estimate for 115/69/34.5-kV Substation

Work Element	Man-hours	Labor Cost	Overhead	Profit	Quote Price
Steel structures	582	$3,724.80	$1,676.16	$ 540.00	$ 5,941.06
Disconnect switches	1,139	7,289.80	3,280.32	1,057.00	11,626.92
Buswork, insulators, connectors	650	4,160.00	1,872.00	603.20	6,635.20
Electric and miscellaneous equipment	216	1,382.40	622.08	200.44	2,204.92
Control house	345	2,380.50	1,071.23	345.17	3,796.90
Conduit, grounding, yard lighting	843	5,395.20	2,427.84	782.30	8,605.34
				Total quote price	$38,810.34

For this estimate, the following data were used:

1. Time to completion — 600 hours
2. Hourly rates
   - Lineman
     - Journeyman — $5.29
     - Foreman — 5.70
     - Truck driver — 4.05
   - Insideman
     - Journeyman — 5.55
     - Foreman — 6.11
3. Insurance and taxes — 0.65
   - Welfare — 0.10
   - Job factor — 0.27
   - Inclement weather — 0.18
   - Mileage allowance — 0.23
4. Overhead as percentage of labor cost — 45%

APPENDIX D. GRAPHIC SYMBOLS FOR ONE-LINE DIAGRAMS

One-line diagrams are a useful way of showing an overall power system arrangement. Table 22 shows the symbols most commonly used in substation layouts. For a complete list of symbols, refer to ANSI Standard Y32.2-1970, available for $11.50 from the American National Standards Institute, Inc., 1430 Broadway, New York, New York 10018.

TABLE 22. Symbols Used To Indicate Substation Layout

Equipment	Basic symbol	Variations	
		Symbol	Definition
Arresters lightning			Arrester plus ground (surge arrester)
			Valve-type arrester
Cable terminations (pothead)			Single-conductor termination
			Three-conductor termination
Capacitor			Capicitor plus ground (surge capacitor)
Capacitor bushing			Capacitor bushing potential device
			Coupling capacitor potential device
coupling capacitor			

TABLE 22. Continued

Equipment	Basic Symbol	Variations Symbol	Definition
Circuit breakers air			Breaker with drawout feature
			Breaker with drawout feature and operating coil
power			Breaker with drawout feature
			Breaker with disconnecting switches
Contact			Normally open (NO)
			Normally closed (NC)
Coil operating			Contact with blowout coil
Fuse			High-voltage primary fuse cutout, dry type; or fuse disconnecting switch
			Same as above
			Drawout mounting
			High-voltage primary fuse cutout, oil type

APPENDIX D. GRAPHIC SYMBOLS FOR ONE-LINE DIAGRAMS 97

TABLE 22. Continued

Equipment	Basic Symbol	Variations	
		Symbol	Definition
Gap protective	→←	→←⊣ɪ	Gap plus ground (surge gap)
Ground	⏚		
Mechanical connection	----- Short dashes connecting equipments	--x--	Mechanical interlock
Key interlock	▫--K--▫		
Electrical interlock	--E--		
Meters and instruments	○		

Letter or letters shall be placed within the circle to indicate the type of instrument:

A	ammeter	RH	varhour meter
D	demand meter	S	synchroscope
F	frequency meter	T	temperature
GD	ground detector	V	voltmeter
MA	milliammeter	VA	volt-ammeter
PF	power-factor	VAR	varmeter
RD	recording demand meter	W	wattmeter
RED	recording	WH	watthour meter

Motor	○	(MOT)	Induction
		(MOT) with coils	Synchronous

TABLE 22. Continued

Equipment	Basic Symbol	Variations Symbol	Definition
Reactor (non-magnetic core)	⟶⦿⦿⦿⟶		
Rectifiers half wave (dry type)	▶︎⊢	◇ (AC/DC bridge)	Full wave (dry type)
power	⊕		
Relay	◯ The relay device or function number should be placed within the circle		
Resistor	⎯▭⎯	⎯▭⎯∥	Grounded
Switch air break	⟋⎯	⎯⎞⎯	Double throw
		⟋⎝	Switch with horn gap
Thermal element	⎯⌒⎯		
Interrupter switch	⊙⎯⎯		

APPENDIX D. GRAPHIC SYMBOLS FOR ONE-LINE DIAGRAMS 99

TABLE 22. Continued

Equipment	Basic Symbol	Variations	
		Symbol	Definition
Double throw disconnecting switch			
Double throw interrupter switch			
3 pole, gary operated disconnect switch			
Fused disconnect switch			
Transformer			Two-winding transformer with taps
			Adjustable mutual inductor, constant-current transformer
			Three-winding transformer
			Autotransformer
			Potential transformer
			Current transformer
			Bushing-type current transformer

TABLE 22. Continued

Equipment	Basic Symbol	Variations	
		Symbol	Definition
Transformer winding connections	The following symbols are used to indicate transformer winding connections and may be placed adjacent to the basic transformer symbol:		
		△	Three-phase three-wire delta
		△̠	Three-phase three-wire delta grounded
		↧△↧	Three-phase four-wire delta grounded
		⅄	Three-phase Y
		⅄̠	Three-phase Y grounded neutral
		⊁	Three-phase zigzag
		⊁̠	Three-phase zigzag grounded
		✳	Six-phase star (or diametrical)
		✳̠	Six-phase star with grounded neutral
		∧	Three-phase open delta
		∧̠	Three-phase open delta grounded at common point

APPENDIX E. SUSPENSION INSULATOR UNITS

TABLE 23. Average Number of Suspension Insulator Units per String for Various Line Voltages

Line voltage	Number of suspension insulator units in string
13,200	2
22,000	2, 3
33,000	2, 3
44,000	3, 4
66,000	4, 5
88,000	5, 6
110,000	6, 7, 8
132,000	8, 9, 10
154,000	9, 10, 11
220,000	12-16
330,000	18-22

From *Lineman's Handbook*, by Edwin B. Kurtz. Copyright 1955. Used by permission of the McGraw-Hill Book Company.

APPENDIX F. CONDUCTOR SPACINGS

TABLE 24. Average Conductor Spacings for the Common Distribution and Transmission Voltages

Line Voltage	Distance between line conductors (in.)
2,300	12-18
6,600	18-24
13,200	18-24
22,000	30-36
33,000	36-48
44,000	48-60
66,000	72
88,000	96
110,000	120
132,000	144
154,000	168
220,000	192
330,000	264

From *Lineman's Handbook*, by Edwin B. Kurtz. Copyright 1955. Used by permission of the McGraw-Hill Book Company.

APPENDIX G. SUBSTATION STANDARDS

TABLE 25. NEMA Substation Standards

Nominal voltage kV rating	7.2	14.4	23	34.5	46	69	115	138	161	230	230	345
Impulse withstand kV	95	110	150	200	250	350	550	650	750	900	1050	1300
Minimum phase spacing for horn gap switches and expulsion power fuses (in.)	36	36	48	60	72	84	120	144	168	192	216	240
Vertical break disconnecting switches, bus supports, and power fuses other than expulsion type recommended phase spacing (in.)	18	24	30	36	48	60	84	96	108	132	156	174
Minimum metal-to-metal distance all disconnecting switches, bus supports and rigid conductors (in.)	7	12	15	18	21	31	53	63	72	89	105	119
Insulator height (in.)	7½	10	12	15	18	29	47	52½	61½	76	90½	105
Minimum clearance to ground for rigid parts (in.)	6	7	10	13	17	25	42	50	58	71	83	104
Recommended and minimum clearance between overhead conductors and ground for personnel safety (ft.)	8	9	10	10	10	11	12	13	14	15	16	18

ABOUT THE AUTHOR

John M. Bifulco is a registered Professional Engineer with more than 15 years of experience in such power utility fields as distribution engineering, substations, relaying and controls, and electrical estimating. He is a graduate of Pennsylvania State University with a Bachelor of Science degree in Electrical Engineering.

He has been a lecturer for a number of electrical engineering courses at Gannon College and has published several articles in *Electrical World*. Mr. Bifulco has served as consultant to a number of firms in the power field in Pennsylvania and Ohio. In the course of his work he helped develop concepts that were patented by his former employer. Patents are pending on several other devices submitted in his own name.

Mr. Bifulco's other professional activities include membership in the Institute of Electrical and Electronic Engineers, the National Society of Professional Engineers, and the Erie (Pennsylvania) Engineering Societies Council.